Lecture Notes in Artificial Intelligence 12547

Subseries of Lecture Notes in Computer Science

Series Editors

Randy Goebel
University of Alberta, Edmonton, Canada
Yuzuru Tanaka
Hokkaido University, Sapporo, Japan
Wolfgang Wahlster
DFKI and Saarland University, Saarbrücken, Germany

Founding Editor

Jörg Siekmann
DFKI and Saarland University, Saarbrücken, Germany

More information about this subseries at http://www.springer.com/series/1244

Matthew E. Taylor · Yang Yu ·
Edith Elkind · Yang Gao (Eds.)

Distributed
Artificial Intelligence

Second International Conference, DAI 2020
Nanjing, China, October 24–27, 2020
Proceedings

 Springer

Editors
Matthew E. Taylor
University of Alberta
Edmonton, AB, Canada

Edith Elkind
University of Oxford
Oxford, UK

Yang Yu
Nanjing University
Nanjing, China

Yang Gao
Nanjing University
Nanjing, China

ISSN 0302-9743   ISSN 1611-3349   (electronic)
Lecture Notes in Artificial Intelligence
ISBN 978-3-030-64095-8   ISBN 978-3-030-64096-5   (eBook)
https://doi.org/10.1007/978-3-030-64096-5

LNCS Sublibrary: SL7 – Artificial Intelligence

This Springer imprint is published by the registered company Springer Nature Switzerland AG
The registered company address is: Gewerbestrasse 11, 6330 Cham, Switzerland

Preface

Lately, there has been tremendous growth in the field of artificial intelligence (AI) in general and in multi-agent systems research in particular. Problems arise where decisions are no longer made by a center but by autonomous and distributed agents. Such decision problems have been recognized as a central research agenda in AI and a fundamental problem in multi-agent research. Resolving these problems requires that different scientific communities interact with each other, calling for collaboration and raising further important interdisciplinary questions. Against this background, a new conference, the International Conference on Distributed Artificial Intelligence (DAI), was organized in 2019. DAI aims at bringing together international researchers and practitioners in related areas including general AI, multi-agent systems, distributed learning, computational game theory, etc., to provide a single, high-profile, internationally renowned forum for research in the theory and practice of distributed AI. The year 2020 was the second year of DAI.

This year, we received 23 valid submissions. Despite the relatively small number of submissions, the authors represent various countries including China, the USA, Singapore, and The Netherlands. Each paper was assigned to three Program Committee (PC) members. We ensured that each paper received at least two reviews and most papers received three reviews. After the review, the program co-chairs went through all the papers again. The final decisions were made based on the discussion and consensus of program co-chairs.

Eventually, 9 out of 23 papers were accepted, resulting in an acceptance rate of 39%. The topics of the accepted papers include reinforcement learning, multi-agent learning, distributed learning systems, deep learning, applications of game theory, multi-robot systems, security games, complexity of algorithms for games, etc. The conference program contained 5 paper oral presentation sessions, including the 9 DAI accepted papers and 13 invited papers on related topics from sister conferences such as IJCAI, AAAI, AAMAS, NeurIPS, ICML, and ICLR. Each presentation was allocated 15 minutes. Moreover, the 9 accepted papers were included in the poster session, together with the 27 related sister conference papers, for in-depth communication between the authors and the audience.

The DAI 2020 conference was conceived at the end of 2019. At that time, no one could foresee that the year 2020 would be so different. The continuing influence of COVID-19 created additional challenges for DAI 2020. In particular, the conference had to be moved online. DAI 2020 would not have be able to continue without the help of many friends. We would like to sincerely thank Bo An, Celine Lanlan Chang, the local organization team, and all PC members for their time, effort, dedication, and services to DAI 2020.

October 2020

Matthew E. Taylor
Yang Yu
Edith Elkind
Yang Gao

Organization

General Chairs

Edith Elkind   University of Oxford, UK
Yang Gao   Nanjing University, China

Program Committee Chairs

Matthew E. Taylor   University of Alberta, Canada
Yang Yu   Nanjing University, China

Program Committee

Bo An   Nanyang Technological University, Singapore
Xiong-Hui Chen   Nanjing University, China
Jianye Hao   Tianjin University and Huawei Noah's Ark Lab, China
Sheng-Yi Jiang   Nanjing University, China
Ziniu Li   The Chinese University of Hong Kong (Shenzhen),
                            China
Jincheng Mei   University of Alberta, Canada
Reuth Mirsky   The University of Texas at Austin, USA
Rongjun Qin   Nanjing University, China
Weiran Shen   Carnegie Mellon University, USA
Zihe Wang   Shanghai University of Finance and Economics, China
Feng Wu   University of Science and Technology of China, China
Lei Yuan   Nanjing University, China
Chongjie Zhang   Tsinghua University, China
Zongzhang Zhang   Nanjing University, China
Dengji Zhao   ShanghaiTech University, China
Ming Zhou   Shanghai Jiao Tong University, China

Contents

Parallel Algorithm for Nash Equilibrium in Multiplayer Stochastic Games with Application to Naval Strategic Planning

Sam Ganzfried[1(✉)], Conner Laughlin[2], and Charles Morefield[2]

[1] Ganzfried Research, Miami Beach, USA
sam.ganzfried@gmail.com
[2] Arctan, Inc., Arlington, USA

Abstract. Many real-world domains contain multiple agents behaving strategically with probabilistic transitions and uncertain (potentially infinite) duration. Such settings can be modeled as stochastic games. While algorithms have been developed for solving (i.e., computing a game-theoretic solution concept such as Nash equilibrium) two-player zero-sum stochastic games, research on algorithms for non-zero-sum and multiplayer stochastic games is limited. We present a new algorithm for these settings, which constitutes the first parallel algorithm for multiplayer stochastic games. We present experimental results on a 4-player stochastic game motivated by a naval strategic planning scenario, showing that our algorithm is able to quickly compute strategies constituting Nash equilibrium up to a very small degree of approximation error.

Keywords: Stochastic game · National security · Nash equilibrium

1 Introduction

Nash equilibrium has emerged as the most compelling solution concept in multiagent strategic interactions. For two-player zero-sum (adversarial) games, a Nash equilibrium can be computed in polynomial time (e.g., by linear programming). This result holds both for simultaneous-move games (often represented as a matrix), and for sequential games of both perfect and imperfect information (often represented as an extensive-form game tree). However, for non-zero-sum and games with 3 or more agents it is PPAD-hard to compute a Nash equilibrium (even for the simultaneous-move case) and widely believed that no efficient algorithms exist [3–5]. For simultaneous (*strategic-form*) games several approaches have been developed with varying degrees of success [1,9,10,12].

This research was developed with funding from the Defense Advanced Research Projects Agency (DARPA). The views, opinions and/or findings expressed are those of the authors and should not be interpreted as representing the official views or policies of the Department of Defense or the U.S. Government.

© Springer Nature Switzerland AG 2020
M. E. Taylor et al. (Eds.): DAI 2020, LNAI 12547, pp. 1–13, 2020.
https://doi.org/10.1007/978-3-030-64096-5_1

While extensive-form game trees can be used to model sequential actions of a known duration (e.g., repeating a simultaneous-move game for a specified number of iterations), they cannot model games of unknown duration, which can potentially contain infinite cycles between states. Such games must be modeled as *stochastic games*.

Definition 1. *A* stochastic game *(aka* Markov game*) is a tuple* (Q, N, A, P, r), *where:*

- *Q is a finite set of (stage) games (aka* game states*)*
- *N is a finite set of n players*
- *$A = A_1 \times \ldots \times A_n$, where A_i is a finite set of actions available to player i*
- *$P : Q \times A \times Q \to [0, 1]$ is the transition probability function; $P(q, a, \hat{q})$ is the probability of transitioning from state q to state $\hat{q}$ after action profile a*
- *$R = r_1, \ldots, r_n$, where $r_i : Q \times A \to \mathbb{R}$ is a real-valued payoff function for player i*

There are two commonly used methods for aggregating the stage game payoffs into an overall payoff: *average (undiscounted) reward* and *future discount reward* using a discount factor $\delta < 1$. Stochastic games generalize several commonly-studied settings, including games with finite interactions, strategic-form games, repeated games, stopping games, and Markov decision problems.

The main solution concept for stochastic games, as for other game classes, is Nash equilibrium (i.e., a strategy profile for all players such that no player can profit by unilaterally deviating), though some works have considered alternative solution concepts such as correlated equilibrium and Stackelberg equilibrium. Before discussing algorithms, we point out that, unlike other classes of games such as strategic and extensive-form, it is not guaranteed that Nash equilibrium exists in general in stochastic games.

One theorem states that if there is a finite number of players and the action sets and the set of states are finite, then a stochastic game with a finite number of stages always has a Nash equilibrium (using both average and discounted reward). Another result shows that this is true for a game with infinitely many stages if total payoff is the discounted sum.

Often a subset of the full set of strategies is singled out called *stationary strategies*. A strategy is *stationary* if it depends only on the current state (and not on the time step). Note that in general a strategy could play different strategies at the same game state at different time steps, and a restriction to stationary strategies results in a massive reduction in the size of the strategy spaces to consider. It has been shown that in two-player discounted stochastic games there exists an equilibrium in stationary policies. For the undiscounted (average-reward) setting, it has recently been proven that each player has a strategy that is ϵ-optimal in the limit as $\epsilon \to 0$, technically called a *uniform equilibrium*, first for two-player zero-sum games [11] and more recently for general-sum games [14].

Thus, overall, the prior results show that for two-player (zero-sum and non-zero-sum) games there exists an equilibrium in stationary strategies for the discounted reward model, and a uniform equilibrium for the average reward model.

However, for more than two players, only the first of these is guaranteed, and it remains an open problem whether a (uniform) equilibrium exists in the undiscounted average-reward model. Perhaps this partially explains the scarcity of research on algorithms for multiplayer stochastic games.

Several stochastic game models have been proposed for national security settings. For example, two-player discounted models of adversarial patrolling have been considered, for which mixed-integer program formulations are solved to compute a Markov stationary Stackelberg equilibrium [15,16]. One work has applied an approach to approximate a correlated equilibrium in a three-player threat prediction game model [2]. However we are not aware of other prior research on settings with more than two players with guarantees on solution quality (or for computing Nash as opposed to Stackelberg or correlated equilibrium).

The only prior research we are aware of for computing Nash equilibria in multiplayer stochastic games has been approaches developed for poker tournaments [7,8]. Our algorithms are based on the approaches developed in that work. The model was a 3-player poker tournament, where each state corresponded to a vector of stack sizes. The game had potentially infinite duration (e.g., if all players continue to fold the game continues forever), and was modeled assuming no discount factor. Several algorithms were provided, with the best-performer being based on integrating fictitious play (FP) with a variant of policy iteration. While the algorithm is not guaranteed to converge, a technique was developed that computes the maximum amount a player could gain by deviating from the computed strategies, and it was verified that in fact this value was quite low, demonstrating that the algorithm successfully computed a very close approximation of Nash equilibrium. In addition to being multiplayer, this model also differed from the previous models in that the stage games had imperfect information.

The main approaches from prior work on multiplayer stochastic game solving integrate algorithms for solving stage games (of imperfect information) assuming specified values for the payoffs of all players at transitions into other stage games, and techniques for updating the values for all players at all states in light of these newly computed strategies. For the stage game equilibrium computation these algorithms used fictitious play, which is an iterative algorithm that has been proven to converge to Nash equilibrium in certain classes of games (two-player zero-sum and certain classes of two-player general-sum). For multiplayer and non-zero-sum games it does not guarantee convergence to equilibrium, and all that can be proven is that if it does happen to converge then the sequence of strategies determined by the iterations constitutes an equilibrium. It did happen to converge consistently in the 3-player application despite the fact that it is not guaranteed to do so, suggesting that it likely performs better in practice than the worst-case theory would dictate. For the value updating step, variants of value iteration and policy iteration (which are approaches for solving Markov decision processes) were used.

In this work we build on the prior algorithms for multiplayer stochastic games to solve a 4-player model of naval strategic planning that we refer to as a *Hostility Game*. This is a novel model of national security that has been devised by a domain expert. The game is motivated by the Freedom of Navigation Scenario in the South China Sea, though we think it is likely also applicable to other situations, and in general that multiplayer stochastic games are fundamental for modeling national security scenarios.

2   Hostility Game

In the South China Sea a set of *blue* players attempts to navigate freely, while a set of *red* players attempt to obstruct this from occurring. In our model there is a single blue player and several red players of different "types" which may have different capabilities (we will specifically focus on the setting where there are three different types of red players). If a blue player and a subset of the red players happen to navigate to the same location, then a confrontation will ensue, which we refer to as a Hostility Game.

In a Hostility Game, each player can initially select from a number of available actions (which is between 7 and 10 for each player). Certain actions for the blue player are *countered* by certain actions of each of the red players, while others are not. Depending on whether the selected actions constitute a counter, there is some probability that the blue player *wins* the confrontation, some probability that the red players win, and some probability that the game repeats. Furthermore, each action of each player has an associated *hostility level*. Initially the game starts in a state of zero hostility, and if it is repeated then the overall hostility level increases by the sum of the hostilities of the selected actions. If the overall hostility level reaches a certain threshold (300), then the game goes into *kinetic mode* and all players achieve a very low payoff (negative 200). If the game ends in a win for the blue player, then the blue player receives a payoff of 100 and the red players receive negative 100 (and vice versa for a red win). Note that the game repeats until either the blue/red players win or the game enters kinetic mode. A subset of the game's actions and parameters are given in Fig. 1. Note that in our model we assume that all red players act independently and do not coordinate their actions. Our game model and parameters were constructed from discussions with a domain expert.

Definition 2. *A hostility game (HG) is a tuple $G = (N, M, c, b^D, b^U, r^D, r^U, \pi, h, K, \pi^K)$, where*

- *N is the set of players. For our initial model we will assume player 1 is a blue player and players 2–4 are red players (P2 is a Warship, P3 is a Security ship, and P4 is an Auxiliary vessel).*
- *$M = \{M_i\}$ is the set of actions, or moves, where M_i is the set of moves available to player i*
- *For $m_i \in M_i$, $c(M_i)$ gives a set of blue moves that are counter moves of m_i*

Move	Vessel	Moves	# of times	Hostility	Probability: Defended	Probability: Undefended
W1	Warship	Bridge to Bridge	1	10	5.00%	7.00%
W2	Warship	Shouldering	Unlimited	25	15.00%	30.00%
W5	Warship	Move towards the Islands	Unlimited	5 + 1 for every lane	5% or 0%	5% or 0%
W6	Warship	Continue on current course	Unlimited	3 + 1 for every lane	5% or 0%	5% or 0%
W7	Warship	K Action	NA	100	45.00%	45.00%
S1	Security Vessel	Bridge to Bridge	1	10	3.00%	5.00%
S2	Security Vessel	Move around AFT end	Unlimited	15	5.00%	15.00%
S4	Security Vessel	Move closer to the ship	Unlimited	55	7.00%	7.00%
S5	Security Vessel	Move towards the Islands	Unlimited	2 + 1 for every lane	3% or 0%	3% or 0%
S6	Security Vessel	Continue on current course	Unlimited	1 + 1 for every lane	3% or 0%	3% or 0%
S7	Security Vessel	K Action	NA	100	25.00%	25.00%
A1	AUX Vessel	Throw out Fishing nets	1	5	3.50%	3.50%
A2	AUX Vessel	Move around AFT end	Unlimited	15	5.00%	5.00%
A3	AUX Vessel	Cross Brow	Unlimited	45	3.00%	3.00%
A4	AUX Vessel	Move closer to the ship	Unlimited	55	7.00%	7.00%
A5	AUX Vessel	Move towards the Islands	Unlimited	2 + 1 for every lane	3% or 0%	3% or 0%
A6	AUX Vessel	Continue on current course	Unlimited	1 + 1 for every lane	3% or 0%	3% or 0%
A7	AUX Vessel	Operates unsafely close to the ship	NA	100	15.00%	15.00%
B1	Blue Ship	Bridge to bridge/Deploy SNOOPIE team	1 For every Vessel interaction	10	W: 5% S&A: 20%	W: 7% S&A: 25%
B2	Blue Ship	Counter Shouldering	Unlimited	30	W: 20% S&A: N/A	N/A
B4	Blue Ship	Continue on current course	Unlimited	2 + 1 for every lane	10.00%	10.00%
B5	Blue Ship	Move towards the Islands	Unlimited	5 + 1 for every lane	0.00%	0.00%
B6	Blue Ship	Speed up	2	15	W: 3% S&A: 25%	W: 3% S&A: 25%
B7	Blue Ship	Alert Crew security	1	60	W: 18% S&A: 75%	W: 18% S&A: 75%
B8	Blue Ship	K Action	NA	100	90.00%	90.00%
B9	Blue Ship	K Action	NA	100	55.00%	55.00%

Fig. 1. Sample of typical actions and parameters for Hostility Game. (Color figure online)

- For each blue player move and red player, a probability of blue success/red failure given that the move is defended against (i.e., countered), denoted as b^D
- Probability that a move is a blue success/red failure given the move is Undefended against, denoted as b^U
- Probability for a red success/blue failure given the move is defended against, r^D
- Probability for a red success/blue failure given the move is not defended against, r^U
- Real valued payoff for success for each player, π_i
- Real-valued hostility level for each move $h(m_i)$
- Positive real-valued kinetic hostility threshold K
- Real-valued payoffs for each player when game goes into Kinetic mode, π_i^K

We model hostility game G as a (4-player) stochastic game with a collection of stage games $\{G_n\}$, where n corresponds to the cumulative sum of hostility levels of actions played so far. The game has $K + 3$ states: $G_0, \ldots, G_K$, with two additional terminal states B and R for blue and red victories. Depending on whether the blue move is countered, there is a probabilistic outcome for whether the blue player or red player (or neither) will outright win. The game will then transition into terminal states B or R with these probabilities, and then will be over with final payoffs. Otherwise, the game transitions into $G_{n'}$ where n' is the new sum of the hostility levels. If the game reaches G_K, the players obtain the kinetic payoff π_i^K. Thus, the game starts at initial state G_0 and after a finite number of time steps will eventually reach one of the terminal states (B, R, G_K).

Note that in our formulation there is a finite number of players (4) as well as a finite number of states $(K + 3)$. Furthermore, with the assumption that hostility levels for all actions are positive, the game must complete within a finite number of stages (because the combined hostility level will ultimately reach K if one of the terminal states B or R is not reached before then). So a Nash equilibrium is guaranteed to exist in stationary strategies, for both the average and discounted reward models. Note that the payoffs are only obtained in the final stage when a terminal state is reached, and so the difference between using average and discounted reward is likely less significant than for games where rewards are frequently accumulated within different time steps.

3 Algorithm

While research on algorithms for stochastic games with more than two players is limited, several prior algorithms have been devised and applied in the context of a poker tournament [7,8]. At a high level these algorithms consist of two different components: first is a *game-solving algorithm* that computes an (approximate) Nash equilibrium at each stage game assuming given values for all players at the other states, and the second is a *value update* procedure that updates values for all players at all states in light of the newly-computed stage-game strategies. For the poker application the stage games were themselves games of imperfect information (the players must select a strategy for every possible set of private cards that they could hold at the given vector of chip stack sizes). The fictitious play algorithm was used for the game-solving step, which applies both to games of perfect and imperfect information. Fictitious play is an iterative self-play algorithm that has been proven to converge to Nash equilibrium in certain classes of games (two-player zero-sum and certain non-zero-sum). For multiplayer and non-zero-sum games it does not guarantee convergence to equilibrium, and all that can be proven is that if it does happen to converge, then the sequence of strategies determined by the iterations constitutes an equilibrium (Theorem 1). It did happen to converge consistently in the 3-player application despite the fact that it is not guaranteed to do so, suggesting that it likely performs better in practice than the worst-case theory would dictate.

In (smoothed) fictitious play each player i plays a best response to the average opponents' strategies thus far, using the following rule at time t to obtain the current strategy,

$$s_i^t = \left(1 - \frac{1}{t}\right) s_i^{t-1} + \frac{1}{t} s_i'^t,$$

where $s_i'^t$ is a best response of player i to the profile s_{-i}^{t-1} of the other players played at time $t-1$ (strategies can be initialized arbitrarily at $t = 0$, and for our experiments we will initialize them to be uniformly random). This algorithm was originally developed as a simple learning model for repeated games, and was proven to converge to a Nash equilibrium in two-player zero-sum games [6]. However, it is not guaranteed to converge in two-player general-sum games or games with more than two players. All that is known is that if it does converge, then the strategies constitute a Nash equilibrium (Theorem 1).

Theorem 1. *[6] Under fictitious play, if the empirical distributions over each player's choices converge, the strategy profile corresponding to the product of these distributions is a Nash equilibrium.*

A meta-algorithm that integrates these two components—stage game solving and value updating—is depicted in Algorithm 1. We initialize the values at all states according to V_0, and alternate between the phase of solving each nonterminal stage game using algorithm A (note that for certain applications it may even make sense to use a different stage game algorithm A_i for different states), and the value update phase using algorithm V. Following prior work we will be using fictitious play for A and variants of value and policy iteration for V, though the meta-algorithm is general enough to allow for alternative choices depending on the setting.

Algorithm 1. Meta-algorithm for multiplayer stochastic game equilibrium computation

---

Inputs: Stochastic game G with set of terminal states $\{T_n\}$ and set of U nonterminal states $\{U_n\}$, algorithm for stage game equilibrium computation A, algorithm for updating values of all nonterminal states for all players V, number of iterations N, initial assignment of state values V_0.

Initialize values for all players for all nonterminal states according to V_0.

for $n = 1$ to N **do**

    for $i = 1$ to U **do**

        Solve stage game defined at U_i using algorithm A assuming values given by V_{n-1}.

        · Let $S_{i,n}$ denote the equilibrium for state i.

    Update the values for all nonterminal states U_i according to algorithm V assuming that strategies $S_{i,n}$ are used at game state U_i.

Output strategies $\{S_{i,N}\}$

The first algorithm previously considered, called VI-FP, instantiates Algorithm 1 using fictitious play for solving stage games and a multiplayer analogue of *value iteration* for updating values [7,8]. As originally implemented (Algorithm 2), the algorithm takes two inputs, which determine the stopping criterion for the two phases. The fictitious play phase halts on a given state when no player can gain more than γ by deviating from the strategies (i.e., the strategies constitute a γ-equilibrium), and the value iteration phase halts when all game state values for all players change by less than δ.

Prior work used a domain-specific initialization for the values V^0 called the Independent Chip Model for poker tournaments [7]. A counterexample was provided showing that VI-FP may actually converge to non-equilibrium strategies if a poor initialization is used [8], and it was suggested based on a prior theorem for value iteration in single-agent Markov decision processes (MDPs) that this phenomenon can only occur if not all values are initialized pessimistically (Theorem 2). We note that there is not a well-defined notion of v^* in our setting, as multiplayer games can contain multiple Nash equilibria yielding different payoffs to the players.

Algorithm 2. VI-FP [8]

Inputs: Degree of desired stage game solution approximation γ, desired max difference between value updates δ

$V^0 = \text{initializeValues}()$
diff $= \infty$
$i = 0$
while diff $> \delta$ **do**
    $i = i + 1$
    regret $= \infty$
    $S = \text{initializeStrategies}()$
    while regret $> \gamma$ **do**
        $S = \text{fictPlay}()$
        regret $= \text{maxRegret}(S)$
    $V^i = \text{getNewValues}(V^{i-1}, S)$
    diff $= \text{maxDev}(V^i, V^{i-1})$
return S

Theorem 2. *[13] In our setting, if v^0 is initialized pessimistically (i.e., $\forall s$, $v^0(s) \le v^*(s)$), value iteration converges (pointwise and monotonically) to v^*.*

We also note that the prior work proposed just one option for a set of halting criteria for fictitious play and value iteration. Since fictitious play is not guaranteed to converge in multiplayer games there is no guarantee that the approximation threshold of γ will be reached for sufficiently small values (and similarly there is no guarantee that a value difference of δ will be obtained for the outer loop). There are several other sensible choices of halting criteria, for example running the algorithms for a specified number of iterations as we have done in our meta-algorithm, Algorithm 1. As we will see when we describe our parallel algorithm, this approach would also allow for more consistency between the runtimes of computations on different cores. Another halting criterion for fictitious play is to run it for a specified number of iterations but output the average strategies that produced lowest approximation error ϵ out of all iterations (not just the final strategies after the last iteration).

The next approach considered by prior work also used fictitious play for the stage-game solving phase but substituted in a variant of the policy-iteration algorithm (Algorithm 4) for value iteration in the value update phase. This algorithm called PI-FP is depicted in Algorithm 3. The new values are computed by solving a system of equations defined by a transition matrix. In effect this corresponds to updating all game state values globally to be consistent with the recently-computed stage game strategies, while the value iteration procedure updates the values locally given the prior values of the adjacent states. Thus, at least intuitively we would likely expect PI-FP to outperform VI-FP for this reason. Unlike VI-FP, for PI-FP it can be proven (Proposition 1) that if the algorithm converges then the resulting strategies constitute a Nash equilibrium (regardless of the initialization). The experimental results of prior work agreed with this intuition, as PI-FP converged to near-equilibrium faster than

---

Algorithm 3. PI-FP [8]

Inputs: Degree of desired stage game solution approximation γ, desired max difference between value updates δ

    $V^0 = $ initializeValues()
    diff $= \infty$
    $i = 0$
    while diff $> \delta$ **do**
        $i = i + 1$
        regret $= \infty$
        $S^0 = $ initializeStrategies()
        while regret $> \gamma$ **do**
            $S^i = $ fictPlay()
            regret $= $ maxRegret(S^i)
        $M^i = $ createTransitionMatrix(S^i)
        $V^i = $ evaluatePolicy(M^i)
        diff $= $ maxDev(V^i, V^{i-1})
    return S^i

VI-FP [8]. This was determined by an ex-post checking procedure to compute the degree of approximation ϵ given by Algorithm 5, with correctness following from Theorem 3 for Algorithm 4. The quantity $v_i^{\pi_i^*, s_{-i}^*}(G_0)$ denotes the value to player i at the initial game state when player i plays π_i^* and his opponents play s_{-i}^*, and $v_i^{s_i^*, s_{-i}^*}(G_0)$ is analogous.

Proposition 1. *If the sequence of strategies $\{s^n\}$ determined by iterations of the outer loop of Algorithm 3 converges, then the final strategy profile s^* is an equilibrium.*

Theorem 3. *[13] Let S be the set of states in M. Suppose S and $A(s)$ are finite. Let $\{v^n\}$ denote the sequence of iterates of Algorithm 4. Then, for some finite N, $v^N = v^*$ and $\pi^N = \pi^*$.*

Proposition 2. *Algorithm 5 computes the largest amount any agent can improve its expected utility by deviating from s^*.*

    The implementations of VI-FP and PI-FP in prior work both used a single core, and involved running fictitious play sequentially at every game state within the stage game update phase. We observe that both of these approaches can be parallelized. Assuming there are $|S|$ states and d cores (and for presentation simplicity assuming that $|S|$ is a multiple of d), we can assign $\frac{|S|}{d}$ of the stage games to each core and run fictitious play independently on d states simultaneously. This will compute equilibrium strategies at all stage games, which can be integrated with the value update phase of both VI-FP and PI-FP. Since the stage game solving phase is the bottleneck step of both algorithms, this parallel algorithm will achieve a linear improvement in runtime by a factor of d. In

---

Algorithm 4. Policy iteration for positive bounded models with expected total-reward criterion

---

1. Set $n = 0$ and initialize the policy π^0 so it has nonnegative expected reward.
2. Let v^n be the solution to the system of equations

$$v(i) = r(i) + \sum_j p_{ij}^{\pi^n} v(j)$$

where $p_{ij}^{\pi^n}$ is the probability of moving from state i to state j under policy π^n. If there are multiple solutions, let v^n be the minimal nonnegative solution.
3. For each state s with action space $A(s)$, set

$$\pi^{n+1}(s) \in \operatorname*{argmax}_{a \in A(s)} \sum_j p_{ij}^a v^n(j),$$

breaking ties so $\pi^{n+1}(s) = \pi^n(s)$ whenever possible.
4. If $\pi^{n+1}(s) = \pi^n(s)$ for all s, stop and set $\pi^* = \pi^n$. Otherwise increment n by 1 and return to Step 2.

Algorithm 5. *Ex post* check procedure

Create MDP M from the strategy profile s^*
Run Algorithm 4 on M (using initial policy $\pi^0 = s^*$) to get π^*
return $\max_{i \in N} \left[v_i^{\pi_i^*, s_{-i}^*}(G_0) - v_i^{s_i^*, s_{-i}^*}(G_0) \right]$

---

addition to incorporating parallelization, our Algorithm 6 differs from the prior approach by allowing for custom stopping conditions for the two phases.

We note that neither VI-FP or PI-FP is guaranteed to converge in this setting (though it has been proven that if PI-FP converges then the resulting strategies constitute a Nash equilibrium [8]). Note that our Hostility Game does not technically fall into the positive bounded model [13], as certain actions can obtain negative payoff. However, the main difference between policy iteration for this model (Algorithm 4) as opposed to the discounted reward model is using the minimal nonnegative solution for Step 2 [13]; for all our experiments the transition matrix had full rank and there was a unique solution, so there is no ambiguity about which solution to use. Furthermore, in a Hostility Game the rewards are only obtained at a terminal state, and the total expected reward is clearly bounded (both in the positive and negative directions). So we can still apply these versions of value and policy iteration to (hopefully) obtain optimal solutions. Note also that for the case where all hostility levels are positive we can guarantee the game will complete within a finite duration and can apply backwards induction; but this will not work in general for the case of zero or negative hostilities where the game has potentially infinite duration, and the stochastic game-solving algorithms will be needed.

Algorithm 6. Parallel PI-FP

Inputs: Stopping condition C_S for stage game solving, stopping condition C_V for value updating, number of cores d

$V^0 = $ initializeValues()

$i = 0$

while C_V not met **do**

    $i = i + 1$

    while C_S not met for each stage game **do**

        Run fictitious play on each stage game on d cores (solving d stage games simultaneously) to obtain S^i

    $M^i = $ createTransitionMatrix(S^i)

    $V^i = $ evaluatePolicy(M^i)

  return S^i

---

4 Experiments

Results for the first 25 iterations of several algorithm variations are given in Fig. 2. All experiments ran the parallel versions of the algorithms with 6 cores on a laptop. The variations include VI-FP and PI-FP with varying numbers of iterations of fictitious play, as well as PI-FP using the version of fictitious play where the strategy with lowest exploitability over all iterations was output (as opposed to the final strategy). We first observe that VI-FP did not converge to equilibrium while all versions of PI-FP did, making PI-FP the clearly preferable choice (note that both converged in prior poker experiments). We also observe that using min exploitability FP led to nearly identical performance as the standard version; since this version also takes longer due to the overhead of having to compute the value of ϵ at every iteration instead of just at the end, we conclude that the standard version of fictitious play is preferable to the version that selects the iteration with minimal exploitability.

For Parallel PI-FP using standard fictitious play, we compared results using 1,000, 5,000, 10,000, 20,000, 25,000, and 50,000 iterations of fictitious play for solving each game state within the inner loop of the algorithm. Each of these versions eventually converged to strategies with relatively low exploitability, with the convergence value of ϵ smaller as more iterations of FP are used. Note that initially we set values for all players at all non-terminal states to be zero, and that the terminal payoffs for a victory/loss are $100/-100$, and for kinetic payoffs are -200 (with $K = 300$); so convergence to $\epsilon = 0.01$ is quite good (this represents 0.01% of the minimum possible payoff of the game). Even just using 1,000 iterations of FP converged to ϵ of around 0.25, which is still relatively small. Note that while the final convergence values were quite low, there was quite a bit of variance in ϵ for the first several iterations, even for the versions with large number of FP iterations (e.g., using 10,000–50,000 iterations spiked up to ϵ exceeding 20 at iteration 6, and using 20,000 and 25,000 spiked up again to ϵ exceeding 25 again at iteration 13). So it is very important to ensure that the algorithm can be run long enough to obtain convergence.

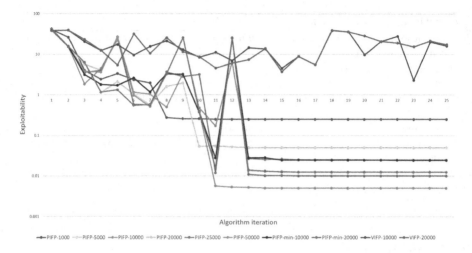

Fig. 2. Performance of several algorithm variants.

5   Conclusion

We have presented a new parallel algorithm for solving multiplayer stochastic games, and presented experimental results showing that it is able to successfully compute an ϵ-equilibrium for very small ϵ for a naval strategic planning scenario that has been devised by a domain expert.

There are several immediate avenues for future study. First, we note that while for the game model we have experimented on the stage games have perfect information, our algorithm also applies to games where the stage games have imperfect information (related prior work has shown successful convergence in the imperfect-information setting for poker tournaments). There are several different natural ways in which imperfect information can be integrated into the model. Currently we are exploring a model in which there is an unknown number of red "sub-players" of each of the three types; this value is known to a single "meta-player" of that type, but the other players only know a publicly-available distribution from which this value is drawn (much like in poker how players receive private cards known only to them and a distribution for the cards of the opponents).

By now we have observed fictitious play to converge consistently for stage games in several domains (previously for poker tournaments and now for naval planning), as well as the general PI-FP algorithm for multiplayer stochastic games. Theoretically we have seen that these approaches are not guaranteed to converge in general for these game classes, and all that has been proven is that if they do converge then the computed strategies constitute a Nash equilibrium (though for VI-FP this is not the case and a counterexample was shown where it can converge to non-equilibrium strategies [8]). It would be interesting from a theoretical perspective to prove more general conditions for which these algorithms are guaranteed to converge in multiplayer settings that can include generalizations of these settings that have been studied.

Many important real-world settings contain multiple players interacting over an unknown duration with probabilistic transitions, and we feel that the multiplayer stochastic game model is fundamental for many national security domains, particularly with the ability of approaches to be integrated with imperfect information and parameter uncertainty. We plan to explore the application of our algorithm to other important domains in the near future.

References

1. Berg, K., Sandholm, T.: Exclusion method for finding Nash equilibrium in multiplayer games. In: Proceedings of the AAAI Conference on Artificial Intelligence (AAAI), pp. 383–389 (2017)
2. Chen, G., Shen, D., Kwan, C., Cruz, J., Kruger, M., Blasch, E.: Game theoretic approach to threat prediction and situation awareness. J. Adv. Inf. Fusion **2**(1), 1–14 (2006)
3. Chen, X., Deng, X.: 3-Nash is PPAD-complete. In: Electronic Colloquium on Computational Complexity Report No. 134, pp. 1–12 (2005)
4. Chen, X., Deng, X.: Settling the complexity of 2-player Nash equilibrium. In: Proceedings of the Annual Symposium on Foundations of Computer Science (FOCS) (2006)
5. Daskalakis, C., Goldberg, P., Papadimitriou, C.: The complexity of computing a Nash equilibrium. SIAM J. Comput. **1**(39), 195–259 (2009)
6. Fudenberg, D., Levine, D.: The Theory of Learning in Games. MIT Press, Cambridge (1998)
7. Ganzfried, S., Sandholm, T.: Computing an approximate jam/fold equilibrium for 3-player no-limit Texas Hold 'em tournaments. In: Proceedings of the International Conference on Autonomous Agents and Multi-Agent Systems (AAMAS) (2008)
8. Ganzfried, S., Sandholm, T.: Computing equilibria in multiplayer stochastic games of imperfect information. In: Proceedings of the 21st International Joint Conference on Artificial Intelligence (IJCAI) (2009)
9. Govindan, S., Wilson, R.: A global Newton method to compute Nash equilibria. J. Econ. Theory **110**, 65–86 (2003)
10. Lemke, C., Howson, J.: Equilibrium points of bimatrix games. J. Soc. Ind. Appl. Math. **12**, 413–423 (1964)
11. Mertens, J.F., Neyman, A.: Stochastic games. Int. J. Game Theory **10**(2), 53–66 (1981). https://doi.org/10.1007/BF01769259
12. Porter, R., Nudelman, E., Shoham, Y.: Simple search methods for finding a Nash equilibrium. Games Econ. Behav. **63**(2), 642–662 (2008)
13. Puterman, M.L.: Markov Decision Processes: Discrete Stochastic Dynamic Programming. Wiley, Hoboken (2005)
14. Vieille, N.: Two-player stochastic games I: a reduction. Israel J. Math. **119**(1), 55–91 (2000). https://doi.org/10.1007/BF02810663
15. Vorobeychik, Y., An, B., Tambe, M., Singh, S.: Computing solutions in infinite-horizon discounted adversarial patrolling games. In: International Conference on Automated Planning and Scheduling (ICAPS) (2014)
16. Vorobeychik, Y., Singh, S.: Computing Stackelberg equilibria in discounted stochastic games. In: Proceedings of the AAAI Conference on Artificial Intelligence (AAAI) (2012)

LAC-Nav: Collision-Free Multiagent Navigation Based on the Local Action Cells

Li Ning$^{(\boxtimes)}$ and Yong Zhang

Shenzhen Institutes of Advanced Technology, CAS, Shenzhen, China
{li.ning,zhangyong}@siat.ac.cn

Abstract. Collision avoidance is one of the most primary problems in the decentralized multiagent navigation: while the agents are moving towards their own targets, attentions should be paid to avoid the collisions with the others. In this paper, we introduced the concept of the *local action cell*, which provides for each agent a set of velocities that are safe to perform. Consequently, as long as the local action cells are updated on time and each agent selects its motion within the corresponding cell, there should be no collision caused. Furthermore, we coupled the local action cell with an adaptive learning framework, in which the performance of selected motions are evaluated and used as the references for making decisions in the following updates. The efficiency of the proposed approaches were demonstrated through the experiments for three commonly considered scenarios, where the comparisons have been made with several well studied strategies.

Keywords: Buffered Voronoi cell · Adaptive learning · Multiagent navigation

1 Introduction

Collision-free navigation is a fundamental and important problem in the design of multiagent systems, which are widely applied in the fields such as robots control and traffic engineering. When moving the agents in an environment with static or dynamic obstacles, it is usually a necessary requirement to well plan the trajectories such that no collision is caused. As the number of agents increases and the environment area becomes large, planning the realtime motions for all agents in a centralized manner may need huge amount of the calculations, and is often restricted by the efficiency of the communication between the agent and the planning monitor. Therefore, it is natural (sometimes necessary) to consider the decentralized navigation approaches, by which the individual agent is responsible for sensing the nearby obstacles and performing the proper motion to progress towards its destination, without causing any collisions. On the other hand, as a consequence of the decentralized navigation, it is in general difficult

© Springer Nature Switzerland AG 2020
M. E. Taylor et al. (Eds.): DAI 2020, LNAI 12547, pp. 14–28, 2020.
https://doi.org/10.1007/978-3-030-64096-5_2

for the agents to fully coordinate before making the independent moves. Thus they should also be considered and avoided as the obstacles to each other.

As noticed in the existing works, when avoiding collisions with the other agents, it is important to take into account the fact that they are also intelligent to perform the collision avoidance motions (otherwise, undesirable oscillations may be observed during the navigation). Consequently, it is not necessary for any individual agent to take all the responsibility of making sure that the performed motion is safe. ORCA [10] is a well known decentralized approach that guarantees to generate the optimal reciprocal collision-free velocities, except for some certain conditions with densely packed agents. BVC [11] has been proposed to restrict the agents moving inside the non-intersecting areas, and thus the collision avoidance is guaranteed. After the safe field of the motions (i.e. the safe range of velocities or the safe area of positions) is determined, both of the ORCA-based approaches and the BVC-based approaches usually select the motion that is closest to the preferred motion, within the safe field. Such a greedy strategy is natural and widely used in the local-search-based optimizations. However, it may cause the less efficient performance in the multiagent navigation, as the agents may refuse to detour until there is no chance to approach the target. In the worst case, with the greedy selection, agents may get stuck in a loop of two or more situations (also known as deadlocks). Although some tricks have been proposed to fix such drawbacks (including the ideas described in [11]), they are not always valid in the concrete implementations, and the improvements vary from case to case.

In this work, in order to improve the navigation efficiency, we extend the buffered Voronoi cell [11] in the velocity space, and consider the relative velocities for their effects on causing the potential conflicts. In the selection of the motion to perform, the traveling progress has been also considered, and consequently the agents may detour earlier, as long as approaching directly to the target leads to less progress in the moving distance.

Problem Formation. In this work, we consider a set $\mathcal{A}$ of the disk-shaped agents moving in the plane. For any time point, agent $a_i \in \mathcal{A}$ of position $p_i \in \mathbb{R}^2$ is free to change its velocity $v_i \in \mathbb{R}^2$, and after a short time $\delta > 0$, it moves to $p_i + \delta \cdot v_i$, if there is no collisions between the agents (i.e. the distance between any pair of agents is at least the sum of their radii). For a decentralized navigation approach, it runs independently for each individual agent a_i, and based on the observations of the environment, it updates the velocity in order to guide agent a_i to arrive at the given and fixed destination/target $d_i \in \mathbb{R}^2$. On the measure of the approach's performance, we want all the agents arriving at their destinations/targets as soon as possible, without causing any collisions.

Our Contributions. We introduced the concept of the *local action cell* to specify the underlying choices for the selection of the motion to perform, and proposed two approaches (LAC-Nav and LAC-Learn) that guarantee to provide the collision-free navigations. While the LAC-Nav approach simply perform the action of the largest penalized length (among all choices in the local action cell), the LAC-Learn approach evaluates the performed actions and adjust the selection based on an adaptive learning framework. The experiment results have shown that the proposed approaches perform more efficiently in the completion time (formally defined in the section of "Experiments"), compared to several well studied approaches.

Related Works. The velocity-based collision-free navigation have been extensively studied in the last two decades. The idea of *reciprocal velocity obstacles* (RVO, [2]) was introduced to reduce the problem of calculating the collision-free motion to solving a low-dimensional linear program, based on the definition of velocity obstacles [4], and it was further improved to derive the *optimal reciprocal collision avoidance* (ORCA, [10]) framework, which guarantees the optimal reciprocal collision-free motions, except for some certain conditions with densely packed agents. While the safety of the final motion is guaranteed by ORCA, the ALAN [6] online learning framework has been proposed for adapting the preferred motions of multiple agents without the need for offline training; and the CNav [7] is designed to allow the agents to take the others' preferred motion into account and adjust accordingly to achieve the better coordination in the crowd environments. Notice that although the efficiency of CNav has been demonstrated through the experiments, it requires the spreading of some private information of the agents, such as their preferred motions or their targets, which is often a controversial issue in the practical applications.

As the well known Voronoi diagram can be used to divide the working space into non-intersecting areas, it has been also adopted for the collision-free path planning with multiple robots [3,5]. Inspired by the algorithms for the coverage control of the agents [8], and a Voronoi-cell-based algorithm [1] which is introduced to avoid collisions within a larger probabilistic swarm, the *buffered Voronoi cell* (BVC, [11]) approach has been proposed to achieve the collision avoidance guarantee for the multiagent navigation, based on only the information of the positions. With the up-to-date information of the others' positions, the agents are restricted to move in the non-intersecting areas, and thus there should be no collisions. In [9], a trajectory planning algorithm was proposed to navigate the agents under the higher-order dynamic limits, in which BVC is used as the low-level strategy to avoid collisions.

2   The Local Action Cells

In this paper, we assume that all the agents in $\mathcal{A}$ have the same radius r for the simplicity of the argument (for the case when the agents have different radii, the arguments in this paper can be directly extended by substituting the classical Voronoi diagram with its weighted variant). Thus for any time and any pair of non-colliding agents a_i and a_j, it always holds that $\|p_{ij}\|_2 \geq 2r$, where p_{ij} stands for $p_j - p_i$.

Recall that in [11], the buffered Voronoi cell of agent a_i is defined as

$$\bar{V}_i = \left\{ p \in \mathbb{R}^2 \mid \left(p - \frac{p_i + p_j}{2} \right) \cdot \frac{p_{ij}}{\|p_{ij}\|_2} + r \leq 0, \forall j \neq i \right\},$$

which implies a safe velocity domain

$$\mathcal{D}_i = \left\{ v \in \mathbb{R}^2 | p_i + \delta \cdot v \in \bar{V}_i \right\},$$

for agent a_i to change and maintain its velocity in order to reach a point in $\bar{V}_i$, where $\delta \in \mathbb{R}^+$ is the length of the time interval between two consecutive updates. Equivalently, domain $\mathcal{D}_i$ can be presented as

$$\mathcal{D}_i = \left\{ v \in \mathbb{R}^2 \mid v \cdot u_{ij} \leq \frac{\|p_{ij}\|_2 - 2r}{2\delta}, \forall j \neq i \right\},$$

where u_{ij} is the unit vector along the same direction with $p_j - p_i$, i.e. $u_{ij} = p_{ij}/\|p_{ij}\|_2$. Obviously, domain $\mathcal{D}_i$ is the intersection of the half-planes P_{ij}'s for each agent $a_j \neq a_i$, with

$$P_{ij} = \left\{ v \in \mathbb{R}^2 \mid v \cdot u_{ij} \leq \frac{\|p_{ij}\|_2 - 2r}{2\delta} \right\}.$$

Assuming that agent a_i is moving at velocity v_i and agent a_j is moving at velocity v_j, we estimate the colliding risk by calculating

$$v_{ij} = \max \left\{ 0, \frac{\|p_{ij}\|_2 - 2r}{2\delta} - v_j \cdot u_{ij} \right\}$$

and

$$\theta_{ij} = \min \left\{ 1, \frac{\|p_{ij}\|_2 - 2r}{v_{ij} \cdot \tau} \right\},$$

and define the *safe half-plane* $\bar{P}_{ij}$ of agent i according to agent a_j as a subset of P_{ij}

$$\bar{P}_{ij} = \left\{ (1 - \lambda + \theta_{ij} \cdot \lambda) \cdot v | v \in P_{ij} \right\},$$

where $\lambda \in [0, 1]$ is the relax factor indicating how much the agent considers the long-sighted decision, and it is set to 0.5 through this paper.

Now, we are ready to define the *local action cell* (LAC) of agent a_i, denoted by $\mathcal{C}_i$, as a subset of velocities in the intersection of all the safe half-planes, i.e.

$$\mathcal{C}_i = \left\{ v \in \cap_j \bar{P}_{ij} \mid \|v\|_2 \leq \min\{v_{max}, \frac{\|d_i - p_i\|_2}{\delta}\}, \right.$$

$$\left. (\rho(v) - \rho(d_i - p_i)) \bmod 2\pi \in \Delta \right\}$$

where v_{max} indicates the maximum moving speed, d_i is the destination/target of agent a_i, $\rho(\cdot)$ denotes the angle (in radians) of the clockwise rotation of the argument vector to align with the positive direction of the x-axis, and Δ is a set of candidate angles which is defined by

$$\Delta = \left\{ k \cdot \frac{\pi}{4} \mid k \in \mathbb{Z}, 0 \leq k < 8 \right\},$$

through this paper. (See Fig. 1 for an illustration of the local action cell of an agent moving through two neighbors.)

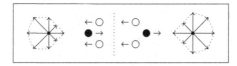

Fig. 1. The local action cell of an agent (the black one) moving through two neighbors.

3   Collision-Free Navigation

In this section, we introduce a distributed approach, named LAC-Nav, for the collision-free navigation with multiple agents. As shown in Algorithm 1, the approach is straight forward with the following steps executed in loops: for each agent a_i, calculate the current local action cell; and then select a proper velocity from the cell.

Algorithm 2 follows the definition of the local action cell and describes the calculation details; Algorithm 3 shows how the new velocity is selected: Given the current local action cell C_i, each velocity $v \in C_i$ is at first evaluated according to the penalized length $\zeta_v \cdot \|v\|_2$, where ζ_v is the factor that is initialized as $0 < \zeta \leq 1$ and decreased exponentially on the angle between v and the direction of $d_i - p_i$. Finally, the velocity of the maximum penalized length is returned as the result.

Algorithm 1: LAC-Nav(a_i): The LAC-based navigation algorithm running on agent a_i.

---

1 **while** a_i *is not at the destination* **do**
2 | $C_i := \mathrm{LAC}(a_i)$;
3 | $v_i^{new} := \mathrm{SelectVel}(C_i)$;
4 | agent a_i moves at velocity v_i^{new};

While calculating the local action cells, it is not necessary to consider all the agents in the environment. When the distance between agent a_i and agent a_j is at least $\ell := 2 \cdot v_{max} \cdot \tau + 2 \cdot r$, it holds directly that $\theta_{ij} = 1$ and $\theta_{ji} = 1$. Thus the corresponding safe half-planes can be ignored in the calculation of the agents' local action cells, which implies it is sufficient to consider only the neighbors within distance ℓ.

Processing Complexity. When considering only the agents within a distance ℓ, the number of an agent's neighbors is at most $3 \cdot \ell^2/r^2$, since there is no overlap between the neighbors and for each of them, at least $1/3$ of the body is covered by the disk of radius ℓ. Consequently, the loop of Lines 6–10 is executed for a constant time within one step of update of an individual agent. Thus, the processing complexity of LAC is determined by the efficiency to detect the neighbors in the specified range. In the simulations, the neighbors can be efficiently

---

Algorithm 2: LAC(a_i): Calculate the current local action cell of agent a_i.

1 $v_i^{max} := \min\{v_{max}, \frac{\|d_i - p_i\|_2}{\delta}\}$;

2 $v_i^0 := v_i^{max} \cdot \frac{d_i - p_i}{\|d_i - p_i\|_2}$;

3 **for** $k = 1$ **to** 7 **do**

4 $\quad\lfloor$ calculate v_i^k such that $\|v_i^k\|_2 = v_i^{max}$ and $\rho(v_i^k) = \big(\rho(v_i^0) + k \cdot \pi/4\big) \bmod 2\pi$;

5 $C_i := \{v_i^k | k \in \mathbb{Z}^*, 0 \le k \le 7\}$;

6 **for** *agent* a_j *with* $j \ne i$ **do**

7 $\quad$ calculate the safe half-plane $\bar{P}_{ij}$;

8 $\quad$ **for** $v \in C_i$ **do**

9 $\quad\quad\lceil$ $\Gamma_v := \{\theta \cdot v \mid \theta \cdot v \in \bar{P}_{ij}, 0 \le \theta \le 1\}$;

10 $\quad\quad\lfloor$ $v := \arg\max_{u \in \Gamma_v} \|u\|_2$;

11 **Return:** C_i;

Algorithm 3: SelectVel(C_i): Select a velocity inside cell C_i as the new velocity to move at.

1 **for** $v \in C_i$ **do**

2 $\quad$ $\alpha_v := (\rho(v) - \rho(d_i - p_i)) \bmod 2\pi$;

3 $\quad$ $\zeta_v := \zeta^{\frac{4\alpha_v}{\pi}}$;

4 $\quad\lfloor$ $w_v := \zeta_v \cdot \|v\|_2$;

5 $v_i^{new} := \arg\max_{v \in C_i} w_v$

6 **Return:** v_i^{new};

---

derived through querying in a KD-Tree that maintains all the positions, and in more practical cases, the neighbor detection is often executed in a parallel process, and it can be assume that the required information is always ready when it is needed.

Learning with LAC. In LAC-Nav, the new velocity is selected according to the penalized length, which can be roughly seen as an estimate of the traveling distance of the next move. On the other hand, it is also common to evaluate the performed actions and record the results, which also provides the information that may be useful for making decisions in the future. In the case when a specific behavior should perform well for a period of time, selecting the action of the best known evaluation should be more promising than trying based on the estimates only. Generally, the evaluations are learned as the agent keeps running in the "sense-evaluate-act" cycles.

Following the ALAN learning framework [6], we propose the LAC-Learn approach, in which the reward of the latest performed action is defined as the summation of the penalized lengths of the velocities in the resulting local action cell. Notice that by this definition, the reward naturally incorporates the considerations of the goal-oriented performance and the politeness performance, which are treated as two separate components in ALAN. In fact, the lengths of the velocities approaching to the destination reflect how efficient the performed action is for getting the agent closer to the goal; and the lengths of velocities in the local action cell as a whole reflects the efficiency in avoiding the crowding situations. In spite of the definition of the action reward, LAC-Learn selects the new velocity in a different way from the one used in ALAN. With LAC-Learn, the selected new velocity is the one corresponding to the action that maximizes a linear combination of the reward and the penalized velocity length.

Algorithm 4: LAC-Learn(a_i): Navigation algorithm of agent a_i while learning with the local action cells.

---

1 **while** a_i *is not at the destination* **do**
2     $C_i := \text{LAC}(a_i)$;
3     $W_i := \text{CalcWeights}(C_i)$;
4     $R_i := \text{UpdateReward}(\alpha_i, W_i, R_i)$;
5     $S_i := \text{UpdateWUCB}(\alpha_i, R_i, S_i)$;
6     $\alpha_i := \text{SelectAct}(\alpha_i, W_i, R_i, S_i)$;
7     $v_i^{new} := C_i[\alpha_i]$;
8   agent a_i moves at velocity v_i^{new};

Algorithm 5: SelectAct(α_i, W_i, R_i, S_i): Select the action for agent a_i to perform.

---

1 $\epsilon_i := 0$;
2 $\alpha := \text{Null}$;
3 **if** $\alpha_i = 0$ **then**
4   **if** $W_i[0] \geq \eta \cdot \min\{v_{max}, \frac{\|d_i - p_i\|_2}{\delta}\}$ **then**
5         $\alpha := 0$;
6         $\epsilon_i := 0$;
7   **else**
8         $\epsilon_i := \min\{1, \epsilon_i + \beta\}$;

9 **if** $\alpha = \text{Null}$ **then**
10   take s from $[0, 1]$ uniformly at random;
11   **if** $s < 1 - \epsilon_i$ **then**
12         $\alpha := \arg\max_{\alpha \in \Delta} ((1 - \gamma) \cdot R_i[\alpha] + \gamma \cdot W_i[\alpha])$;
13   **else**
14         $\alpha := \arg\max_{\alpha \in \Delta} S_i[\alpha]$;

15 **Return:** α;

Inside an execution cycle of some agent a_i, after the local action cell is calculated by LAC (Algorithm 2), the penalized length of each velocity in C_i is calculated as what has been done in Line 2–4 of SelectVel (Algorithms 3), and saved in a set W_i. In UpdateReward, the reward of the last performed action is updated to the sum of all weights in W_i, as mentioned earlier.

Notice that although the velocities given by C_i may vary from step to step, in the local view, they can always be interpreted as the actions corresponding to the angles specified in Δ. For example, without considering the variation of the length, the velocity pointing to the destination can always be interpreted as the action corresponding to angle $0 \in \Delta$. For an action/angle $\alpha \in \Delta$, we use $C_i[\alpha]$ to denote the velocity $v \in C_i$ such that $(\rho(v) - \rho(d_i - p_i)) \bmod 2\pi = \alpha$.

Following the ALAN learning framework, we calculate (by UpdateWUCB) and maintain (in S_i) the upper confidence bound within a moving time window (i.e. a sequence of consecutive time steps), which is used when the agent explores in the action space. As defined in [6], the wUCB score of action α during the last $T \in \mathbb{Z}^+$ steps is defined by

$$\text{wUCB}(\alpha) := \bar{R}_i(\alpha) + \sqrt{\frac{2\ln(\nu)}{\nu_\alpha}},$$

where $\bar{R}_i(\alpha)$ is the average reward of action α, ν_α denotes the number of times action α has been chosen, and ν denotes the total number of performed actions, all with respect to the moving time window.

Similar to the context-aware action selection approach proposed in [6], SelectAct (Algorithm 5) decides with the "win-stay, lose-shift" strategy and the adaptive ϵ-greedy strategy in which the wUCB suggested action is chosen for the exploration.

When the agent is in the winning state (i.e. the goal-oriented action $\alpha_i = 0$ is performed in the last update and is still a good choice in the sense that the corresponding velocity is little constrained), it is natural to keep forwarding to the goal. Otherwise, if the agent is in the losing state, it performs the ϵ-greedy strategy to exploit on the action that maximizes a linear combination of the action reward and the penalized length of the corresponding velocity. With a small and adaptively adjusted probability, the agent explores and performs the action that maximizes the wUCB score.

The hyper-parameters $\eta \in [0,1]$ (Line 4 in Algorithm 5) and $\gamma \in [0,1]$ (Line 12 in Algorithm 5) are determined depends on the scenarios.

4 Experiments

In this section, we present the results of running experiments with LAC-Nav and LAC-Learn, on a computer of 7 Intel Core i7-6700 CPU (3.40 GHz) processors. The simulations are implemented in Python 3.5, while the update processes of individual agents have been speeded up by applying the multitasking scheme. For one second, there are 100 updates performed for each agent, and therefore we set $\delta := 0.01$ in the implementation of LAC-Nav and LAC-Learn.

Scenarios. For the experiments, we considered three scenarios (Fig. 2): the *reflection* scenario, the *circle* scenario and the *crowd* scenario, where

- in the reflection scenario, two groups of agents start from the left side and right side of the area, respectively (Fig. 2(a)). For each agent, the target is the position on the other side that is symmetric to its start position (Fig. 2(d)). Through navigating the agents to the target positions, the picture of the starting configuration is reflected.
- in the circle scenario, the agents start in layers of circles (Fig. 2(b)), and each agent targets the antipodal position (Fig. 2(e)). That is, the picture of starting configuration is going to be "rotated" by half of a circle, around the origin/center.
- in the crowd scenario, the start positions (Fig. 2(c)) and target positions (Fig. 2(f)) are randomly picked from a small area.

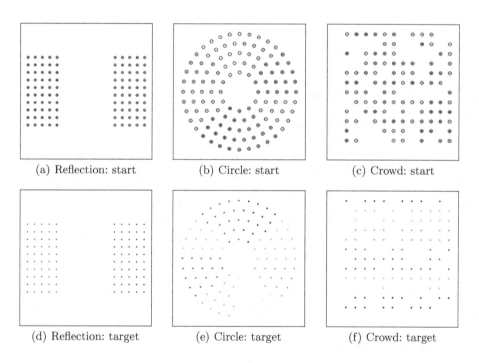

(a) Reflection: start   (b) Circle: start   (c) Crowd: start

(d) Reflection: target   (e) Circle: target   (f) Crowd: target

Fig. 2. Experiment scenarios.

For each of these three scenarios, we compare the performances of LAC-Nav and LAC-Learn, with the performances of approaches including BVC [11], CNav [7], ALAN [6] and ORCA [10].

For the performance evaluation of the considered approaches, we consider three measurements in this work: the *completion time*, the *average detour-distance ratio*, and the *average detour-time ratio*, where

- the completion time of running a navigation algorithm is defined as the time (in seconds) when the last agent arrives at its target, assuming all the agents start from time 0;
- the average detour-distance ratio is defined as the average of the ratios between the actual travel distance and the optimal travel distance (i.e. the length of the straight line from the start position to the target position), over all the agents.
- the average detour-time ratio is defined as the average of the ratios between the actual travel time and the optimal travel time (i.e. the time of moving in a straight line from the start position to the target position, at the maximum speed), over all the agents.

While the completion time justifies the algorithm's global performance on finishing the navigation tasks, by investigating the detour-distance/time ratio, it provides a view on the variance of the individual agent's behavior with different algorithms.

Selection of the Parameters[1]. In the experiments for all scenarios, the agent's radius is uniformly set as $r = 10$, and the maximum moving speed is set as $v_{max} = 50$. In addition, as mentioned in the beginning of this section, within each second there are 100 updates performed for each of the agents, which implies the that the time interval between two consecutive updates is 0.01, i.e. $\delta = 0.01$ in all the experiments.

Recall that when calculating the local action cells, the hyper-parameter $\tau > 0$ is needed to locate the safe half-planes. Through all experiments involving the local action cells, we set $\tau = 0.05$. In addition, the penalty factor ζ is also needed in the calculation of the local action cells (thus it is required when running LAC-Nav and LAC-Learn). Through the experiments, we set $\zeta = 0.95$, and accordingly the goal-orthogonal action (with angle $\pi/2$ from the direction pointing to the goal) is penalized by 0.9025, and the goal-opposite action (along the direction leaving the goal) is penalized by about 0.8145.

For LAC-Learn, we set the mixing factor (Line 12 of Algorithm 5) as $\gamma = 0.75$ for the reflection scenario and the circle scenario, and $\gamma = 0.95$ for the crowd scenario. and the length of the moving time window (for the calculation of wUCB) as $T = 8$, which is the minimum choice as there are 8 actions in the used Δ. Furthermore, the incremental step (for adjusting the exploration probability) is set as $\beta = 0.1$ (Line 8 of Algorithm 5).

---

[1] In general, for the parameters that should be specified in the experiments, we tested with several values, including the one recommended in the paper that proposed the considered algorithm, and selected the best choice.

For ORCA, the collision-free time window is set as $\tau = 0.02$, i.e. twice of the update interval's length. Recall that with CNav and ALAN, ORCA is also called to make sure the performed velocity is collision-free, where the time window are also set as $\tau = 0.02$. Notice that the time window $\tau > 0$ in ORCA has different meaning from the hyper-parameter using the same symbol in the calculation of the local action cells, even though they are both related to the avoidance of the potential collisions.

For CNav, the hyper-parameter for mixing the goal-oriented reward and the constrained-reduction reward is set as $\gamma = 0.5$ for the reflection scenario, and $\gamma = 0.9$ for the circle scenario and the crowd scenario; the number of constrained neighbors of which the action's effect is estimated is set as $k = 3$; the number of the neighbor-based actions is set as $s = 3$.

For ALAN, the hyper-parameter for mixing the goal-oriented reward and the politeness reward is set as $\gamma = 0.5$; the length of the moving time window for the calculation of wUCB is set as $T = 5$; and the incremental step for adjusting the exploration probability is set as $\beta = 0.1$.

Results. In Fig. 3, it shows the experiment results for

- the reflection scenario of 100 agents (50 agents on each side);
- the circle scenario of 120 agents (in 5 circles around the same center point);
- the crowd scenario of 100 agents located in the area of size 600×600.

Overall, LAC-Nav and LAC-Learn outperform almost all the other approaches in the completion time. The only exception is in the reflection scenario, BVC has shown the advantages and it completes earlier than LAC-Learn.

In general, the efficiency of the LAC based approaches is due to the fact that it considers both of the task to arrive at the target and the intension to move as much as possible in every step. The later consideration prevents the agent from the non-necessary halting before it arrives at the target. By maintaining and comparing the penalized lengths of all the candidate actions (according to Δ), even though the agent still has the change to move directly towards the target, it detours as long as there is an other action that provides a better moving (penalized) velocity. As shown in the reflection scenario, this kind of active detouring results in a more fluent navigation as the agents (of the antagonistic moving directions) pass by each other (Fig. 4).

Recall that according to the definition of the safe half-planes, the local action cell is depressed if there are neighbors approaching. Therefore, with the same relative position, it is easier for an agent to "follow" a leaving-away neighbor, if they have the similar preferred trajectories. As a consequence result, in the case when there are more conflicts, such as the circle scenario, after gathered around the central area, instead of squeezing through (as what happens with the other approaches), the agents with the LAC based approaches spin as a whole to resolve the conflicts (Fig. 5).

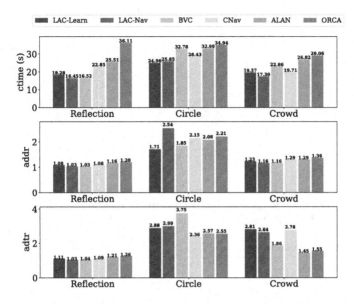

Fig. 3. Experiment results, where *ctime* (s) stands for the completion time (in seconds); *addr* stands for the average detour-distance ratio; and *adtr* stands for the average detour-time ratio.

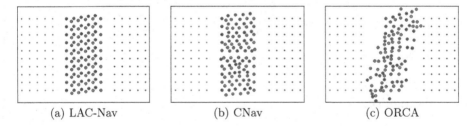

(a) LAC-Nav (b) CNav (c) ORCA

Fig. 4. The antagonistic agents pass by each other in the reflection scenario, where the points of black outline and colored (red/blue) inside are the current positions of the agents, and the simply colored (red/blue) points are the target positions of the agents. (Color figure online)

Although the local action cell can be seen as a variant or extension of the buffered Voronoi cell, it should be noticed that the LAC based approaches perform distinguishably from BVC, except for the simple situation such as the reflection scenario. For the more crowding situations (like the circle scenario and the crowd scenario), the individual agents with LAC-Nav or LAC-Learn spend more time on average, while on the other hand the global completion time is shorter. By investigating into the experiment processes, it can be found that the LAC based approaches caused less stuck agents than the other approaches did. This fact can also be revealed by checking the completion time of the first 90% arrivals (Fig. 6), in which the approaches' performances are less distinguishable.

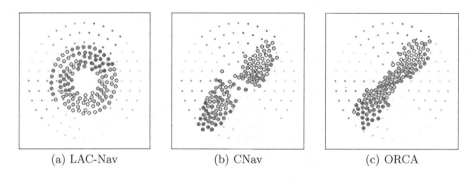

(a) LAC-Nav   (b) CNav   (c) ORCA

Fig. 5. Agents resolve the conflicts in the circle scenario, where the points of black outline and colored inside are the current positions of the agents, and the simply colored points are the target positions of the agents. (Color figure online)

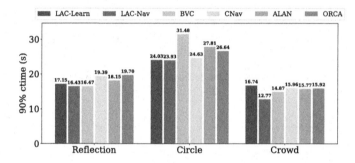

Fig. 6. The completion time (in seconds) of the first 90% arrivals.

5 Discussions

In this work, we introduced the definition of the local action cells, and proposed two approaches LAC-Nav and LAC-Learn, of which the efficiency in the completion time have been experimentally demonstrated. In order to improve the approaches' performance, besides trying with different parameter values, there are some natural directions that also extend the proposed approaches or make a variant.

Adaptive λ. Recall that in the definition of the safe half-plane, we have set the relax factor as $\lambda := 0.5$. Intuitively, λ indicates how much the agents would like to compromise in the next move, in order to avoid the collisions that may happen in a near future. Although it is valid to select any value $[0, 1]$ from the theoretical respect, it should be noted that a very small λ may cause the local action cell being depressed too much, and a very large λ may help little for the long-sighted consideration. The value 0.5 is a balanced choice, and it also follows from an important idea in the reciprocal collision avoidance: each agent take half of the responsibility to avoid the coming collisions. However, it will be more interesting if λ can be dynamic adjusted as the agents learned more information about the environment.

Continuous LAC. In this paper, we defined the local action cells as sets of finite number of actions. However, it may be more natural to consider the continuous area spanned by the velocities in a cell. There is a direct way to extend the definition of the local action cell to include all the linear combinations between every pair of the adjacent velocities. Formally speaking, we can define,

$$C_i^* = \{\lambda_v \cdot v + \lambda_u \cdot u \mid v \text{ and } u \text{ are adjacent in } C_i,$$
$$\lambda_v + \lambda_u \leq 1, \lambda_v \geq 0, \lambda_u \geq 0\},$$

which is a continuous area in the velocity space. With the continuous version the local action cell, the agents are no longer restricted to select the actions from Δ, and they can move in any angle as long as the corresponding velocity has a positive length.

References

1. Bandyopadhyay, S., Chung, S., Hadaegh, F.Y.: Probabilistic swarm guidance using optimal transport. In: 2014 IEEE Conference on Control Applications (CCA), pp. 498–505, October 2014. https://doi.org/10.1109/CCA.2014.6981395
2. van den Berg, J.P., Lin, M.C., Manocha, D.: Reciprocal velocity obstacles for real-time multi-agent navigation. In: 2008 IEEE International Conference on Robotics and Automation, pp. 1928–1935 (2008)
3. Bhattacharya, P., Gavrilova, M.L.: Roadmap-based path planning - using the Voronoi diagram for a clearance-based shortest path. IEEE Robot. Autom. Mag. 15(2), 58–66 (2008). https://doi.org/10.1109/MRA.2008.921540
4. Fiorini, P., Shiller, Z.: Motion planning in dynamic environments using velocity obstacles. Int. J. Robot. Res. 17(7), 760–772 (1998). https://doi.org/10.1177/027836499801700706
5. Garrido, S., Moreno, L., Blanco, D.: Voronoi diagram and fast marching applied to path planning, vol. 2006, pp. 3049–3054, January 2006. https://doi.org/10.1109/ROBOT.2006.1642165
6. Godoy, J.E., Karamouzas, I., Guy, S.J., Gini, M.: Adaptive learning for multi-agent navigation. In: Proceedings of the 2015 International Conference on Autonomous Agents and Multiagent Systems, AAMAS 2015, pp. 1577–1585 (2015)
7. Godoy, J.E., Karamouzas, I., Guy, S.J., Gini, M.L.: Implicit coordination in crowded multi-agent navigation. In: Proceedings of 17th AAAI Conference on Artificial Intelligence, AAAI 2016, pp. 2487–2493 (2016)
8. Pimenta, L.C.A., Kumar, V., Mesquita, R.C., Pereira, G.A.S.: Sensing and coverage for a network of heterogeneous robots. In: 2008 47th IEEE Conference on Decision and Control, pp. 3947–3952, December 2008. https://doi.org/10.1109/CDC.2008.4739194
9. Şenbaşlar, B., Hönig, W., Ayanian, N.: Robust trajectory execution for multi-robot teams using distributed real-time replanning. In: Correll, N., Schwager, M., Otte, M. (eds.) Distributed Autonomous Robotic Systems. SPAR, vol. 9, pp. 167–181. Springer, Cham (2019). https://doi.org/10.1007/978-3-030-05816-6_12

10. van den Berg, J., Guy, S.J., Lin, M., Manocha, D.: Reciprocal n-body collision avoidance. In: Pradalier, C., Siegwart, R., Hirzinger, G. (eds.) Robotics Research. STAR, vol. 70, pp. 3–19. Springer, Heidelberg (2011). https://doi.org/10.1007/978-3-642-19457-3_1
11. Zhou, D., Wang, Z., Bandyopadhyay, S., Schwager, M.: Fast, on-line collision avoidance for dynamic vehicles using buffered Voronoi cells. IEEE Robot. Autom. Lett. **2**(2), 1047–1054 (2017). https://doi.org/10.1109/LRA.2017.2656241

MGHRL: Meta Goal-Generation for Hierarchical Reinforcement Learning

Haotian Fu[1(✉)], Hongyao Tang[1], Jianye Hao[1,2], Wulong Liu[2], and Chen Chen[2]

[1] Tianjin University, Tianjin, China
{haotianfu,bluecontra,jianye.hao}@tju.edu.cn
[2] Noah's Ark Lab, Huawei, Beijing, China
{liuwulong,chenchen9}@huawei.com

Abstract. Most meta reinforcement learning (meta-RL) methods learn to adapt to new tasks by directly optimizing the parameters of policies over primitive action space. Such algorithms work well in tasks with relatively slight differences. However, when the task distribution becomes wider, it would be quite inefficient to directly learn such a meta-policy. In this paper, we propose a new meta-RL algorithm called Meta Goal-generation for Hierarchical RL (MGHRL). Instead of directly generating policies over primitive action space for new tasks, MGHRL learns to generate high-level meta strategies over subgoals given past experience and leaves the rest of how to achieve subgoals as independent RL subtasks. Our empirical results on several challenging simulated robotics environments show that our method enables more efficient and generalized meta-learning from past experience and outperforms state-of-the-art meta-RL and Hierarchical-RL methods in sparse reward settings.

Keywords: Deep Reinforcement Learning · Meta learning · Hierarchical reinforcement learning

1 Introduction

Deep Reinforcement Learning (DRL) has recently shown great success on a wide range of tasks, ranging from games [16] to robotics control [5,13]. However, for more complex problems with larger state and action spaces or sparse reward settings, traditional DRL methods hardly works. Hierarchical reinforcement learning (HRL) in which multiple layers of policies are trained to learn to operate on different levels of temporal abstraction, has long held the promise to learn such difficult tasks [3,6,18]. By decomposing a complex problem into subproblems, HRL significantly reduces the difficulty of solving specific task. Learning multiple levels of policies in parallel is challenging due to non-stationary state transition functions. Recent HRL approaches [14,17] use states as goals directly, allowing simple and fast training of the lower layer.

Human intelligence is remarkable for their fast adaptation to many new situations using the knowledge learned from past experience. However, agents trained

© Springer Nature Switzerland AG 2020
M. E. Taylor et al. (Eds.): DAI 2020, LNAI 12547, pp. 29–39, 2020.
https://doi.org/10.1007/978-3-030-64096-5_3

by conventional DRL methods mentioned above can only learn one separate policy per task, failing to generalize to new tasks without additional large amount of training data. Meta reinforcement learning (meta-RL) addresses such problems by learning how to learn. Given a number of tasks with similar structures, meta-RL methods enable agents to learn such structure from previous experience on many tasks. Thus when encountering a new task, agents can quickly adapt to it with only a small amount of experience.

Most current meta-RL methods leverage experience from previous tasks to adapt to new tasks by directly learning the policy parameters over primitive action space [9,20]. Such approaches suffer from two problems: (i) For complex tasks which require sophisticated control strategies, it would be quite inefficient to directly learn such policy with one nonlinear function approximator and the adaptation to new tasks is prone to be inaccurate. This problem can become more severe in sparse reward settings. (ii) Current meta-RL methods focus on tasks with narrow distribution, how to generalize to new tasks with much more difference remains a problem.

In this paper, we aim at tackling the problems mentioned above by proposing a new hierarchical meta-RL method that meta-learns high-level goal generation and leaves the learning of low-level policy for independent RL. Intuitively, this is quite similar to how a human being behaves: we usually transfer the overall understanding of similar tasks rather than remember specific actions. Our meta goal-generation framework is built on top of the architecture of PEARL [20] and a two level hierarchy inspired by HAC [14]. Our evaluation on several simulated robotics tasks [19] as well as some human-engineered wider-distribution tasks shows the superiority of MGHRL to state-of-the-art meta-RL method.

2   Related Work

Our algorithm is based on meta learning framework [4,24,27], which aims to learn models that can adapt quickly to new tasks. Meta learning algorithms for few-shot supervised learning problems have explored a wide variety of approaches and architectures [21,23,30]. In the context of reinforcement learning, recurrent [8,31] and recursive [15] meta-RL methods adapt to new tasks by aggregating experience into a latent representation on which the policy is conditioned. Another set of methods is gradient-based meta reinforcement learning [9,22,25,32]. Its objective is to learn an initialization such that after one or few steps of policy gradients the agent attains full performance on a new task. These methods focus on on-policy meta learning which are usually sample inefficient. Our algorithm is closely related to probabilistic embeddings for actor-critic RL (PEARL) [20], which is an off-policy meta RL algorithm. PEARL leverages posterior sampling to decouple the problems of inferring the task and solving it, which greatly enhances meta-learning efficiency. However, when facing complex tasks that require sophisticated control strategies, PEARL cannot effectively learn a proper meta-policy as we will show in Sect. 5.

Discovering meaningful and effective hierarchical policies is a longstanding research problem in RL [2,6,7,18,26]. Schmidhuber [24] proposed a HRL approach that can support multiple levels. Multi-level hierarchies have the potential to accelerate learning in sparse reward tasks because they can divide a problem into a set of short-horizon subproblems. Nachum et al. [17] proposed HIRO, a 2-level HRL approach that can learn off-policy and outperforms two other popular HRL techniques used in continuous domains: Option-Critic [2] and FeUdal Networks [29]. Our algorithm is built on Hierarchical actor-critic [14], which is a framework that can learn multiple levels of policies in parallel. Most current HRL works focus on the learning problem in a single task and few of them considers to take advantage of HRL for multi-task or meta-learning tasks. MLSH [10] is such a work which also combines meta-RL with Hierarchical RL. It focuses on meta learning on the low level policy and needs to retrain its high-level policy when facing new tasks. In contrast, with the key insight that humans leverage abstracted prior knowledge obtained from past experience, our method focuses on meta learning high level overall strategy using past experience and leave the detailed action execution for independent RL.

3 Preliminaries

In our meta learning scenario, we assume a distribution of tasks $p(\tau)$ that we want our model to adapt to. Each task correspond to a different Markov Decision Process (MDP), $M_i = \{S, A, T_i, R_i\}$, with state space S, action space A, transition distribution T_i, and reward function R_i. We assume that the transitions and reward function vary across tasks. Meta-RL aims to learn a policy that can adapt to maximize the expected reward for novel tasks from $p(\tau)$ as efficiently as possible.

PEARL [20] is an off-policy meta-reinforcement learning method that drastically improves sample efficiency comparing to previous meta-RL algorithms. The meta-training process of PEARL learns a policy that adapts to the task at hand by conditioning the history of past transitions, which we refer to as context c. Specifically, for the ith transition in task τ, $c_i^\tau = (s_i, a_i, r_i, s_i')$. PEARL leverages an inference network $q_\phi(z|c)$ and outputs probabilistic latent variable z. The parameters of $q(z|c)$ are optimized jointly with the parameters of the actor $\pi_\theta(a|s, z)$ and critic $Q_\theta^h(s, a, z)$, using the reparametrization trick [12] to compute gradients for parameters of $q_\phi(z|c)$ through sampled probabilistic latent variable z.

4 Algorithm

4.1 Two-Level Hierarchy

We set up a hierarchical two-layer RL structure similar to HAC. The high level network uses policy μ^h to generate goals for temporally extended periods in terms of desired observations. In our task, they correspond to the positional

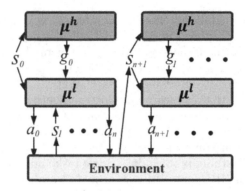

Fig. 1. Two-level hierarchy. High level policy μ^h takes in state and outputs goals at intervals. Low level policy μ^l takes in state and desired goals to generate primitive actions.

features of the gripper. The low level policy μ^l directly controls the agent and produces actions for moving towards the desired goals.

As shown in Fig. 1, the high level policy μ^h observes the state and produces a high level action (or goal) g^t. Low level policy μ^l has at most K attempts of primitive action to achieve g^t. Here, K which can be viewed as the maximum horizon of a subgoal action is a hyperparameter given by the user. As long as the low level policy μ^l run out of K attempts or g^t is achieved, this high level transition terminates. The high level policy uses agent's current state as the new observation and produced another goal for low level policy to achieve.

We use an intrinsic reward function in which a reward of 0 is granted only if the goal produced by high level policy is achieved and a reward of -1 otherwise. Note that the environment's return (i.e. whether the agent successfully accomplished the task) will not affect the reward received by the low level policy. In our evaluation on simulated robotics environments, we use the positional features of the observations as the representation for g^t. A goal g^t is judged to be achieved only if the distance between g^t and the gripper's current position s_{n+1} is less than threshold l.

4.2   Meta Goal-Generation for Hierarchical Reinforcement Learning

The primary motivation for our hierarchical meta reinforcement learning strategy is that, when people try to solve new tasks using prior experience, they usually focus on the overall strategy we used in previous tasks instead of the primitive action execution mechanism. For instance, when we try to use the knowledge learned from riding bicycle to accelerate learning for riding motorcycle, the primitive action execution mechanism is entirely different although they share a similar high-level strategy (e.g. learn how to keep balance first). Thus, we take advantage of our two-level hierarchy structure and propose a new meta-RL framework called meta goal-generation for hierarchical RL (MGHRL). Instead

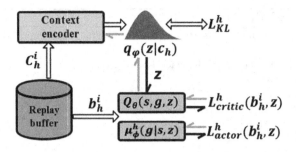

Fig. 2. High level meta-training framework of MGHRL. The context encoder network uses high-level context data C_h^i to infer the posterior over the latent context variable z, and is optimized with gradients from the critic as well as from an information bottleneck on z. The actor network $\mu_\phi^h(g|s,z)$ and critic network $Q_\theta(s,g,z)$ both treat z as part of the state.

of learning to generate a detailed strategy for new tasks, MGHRL learns to generate overall strategy (subgoals) given past experience and leaves the detailed method of how to achieve the subgoals for independent RL. We leverage PEARL framework [20] to independently train a high level meta-policy which is able to quickly adapt to new tasks and generate proper subgoals. Note that off-policy RL method is indispensable in our structure when training high level policy due to its excellent sample efficiency during meta-training. And its structured exploration by reasoning about uncertainty over tasks is crucial to hierarchical parallel training framework. We leave the low level policy to be trained independently with non-meta RL algorithm using hindsight experience replay mechanism [1]. In our simulated robotics experiments, the low level policy aims to move the gripper to the desired subgoal position which can be reused when switching to other tasks. Thus we only need to train a single set of low-level polices which can be shared across different tasks. In other situations where the tasks are from different domains, we can choose to train low level policy independently on new tasks without using past experience.

We summarize our meta-training procedure in Algorithm 1 and Fig. 2. For each training task drawn from task distribution, we sample context c_h and generate hindsight transitions[1] for both levels of hierarchy (*line* 4–13) by performing current policy. Then we train high level and low level networks with the collected data (*line* 16–22).

---

[1] To achieve parallel training for the two levels of our framework, we rewrite past experience transitions as hindsight action transitions, and supplement both levels with additional sets of transitions as was done in HAC.

Algorithm 1. MGHRL Meta-training

Require: Batch of training tasks $\{\tau_i\}_{i=1,...,T}$ from $p(\tau)$, maximum horizon K of sub-goal action

1: Initialize replay buffers $\mathcal{B}_h^i$,$\mathcal{B}_l^i$ for each training task
2: **while** not done **do**
3:   **for** each task τ_i **do**
4:     Initialize high-level context $c_h^i = \{\}$
5:     **for** m=1,...,M **do**
6:       Sample $z \sim q_\phi(z|c_h^i)$
7:       $g_i \leftarrow \mu_h(g|s,z)$
8:       **for** K attempts or until g_i achieved **do**
9:         Gather data using $a_i \leftarrow \mu_l(a|s,g)$
10:         Generate hindsight action transition, hindsight goal transition and add
         to $\mathcal{B}_l^i$
11:       **end for**
12:       Generate hindsight transitions, subgoal test transitions and add to $\mathcal{B}_h^i$
13:       Sample high level context $c_h^i = \{s_j, g_j, r_j, s_j'\}_{j=1,...,N} \sim \mathcal{B}_h^i$
14:     **end for**
15:   **end for**
16:   **for** each training step **do**
17:     **for** each task τ_i **do**
18:       Sample high level context $c_h^i \sim \mathcal{B}_h^i$ and RL batch $b_h^i \sim \mathcal{B}_h^i$, $b_l^i \sim \mathcal{B}_l^i$
19:       Sample $z \sim q_\phi(z|c_h^i)$ and calculate $L_{actor}^h(b_h^i, z)$, $L_{critic}^h(b_h^i, z)$, L_{KL}^h
20:       Update low level actor and critic network with b_l^i
21:     **end for**
22:     Update high level networks with $\sum_i L_{actor}^h$, $\sum_i L_{critic}^h$, $\sum_i L_{KL}^h$
23:   **end for**
24: **end while**

---

5   Experiments

5.1   Environmental Setup

We evaluated our algorithm on several challenging continuous control robotics tasks (integrated with OpenAI Gym), simulated via the MuJoCo simulator [28]:

Fetch-Reach Fetch has to move the gripper to the desired goal position. This task is very easy to learn and is therefore a suitable benchmark to ensure that a new idea works at all.

Fetch-Push A box is placed on a table in front of the robot and Fetch has to move a box by pushing it until it reaches a desired goal position. The robot fingers are locked to prevent grasping. The learned behavior is usually a mixture of pushing and rolling.

Fetch-Slide A puck is placed on a long slippery table and the target position is outside of the robot's reach so Fetch has to hit the puck across a long table such that it slides and comes to rest on the desired goal.

Fetch-PickandPlace Fetch has to pick up a box from a table using its gripper and move it to a desired goal located on the table.

We compare our algorithm to baselines including PEARL with dense reward, HER-PEARL with sparse reward and HAC with shared policy. Note that Rakelly et al. [20] has already shown that PEARL greatly outperforms other existing meta-RL methods like MAML [9], ProMP [22] at both sample efficiency and final performance. Thus we mainly compare our results with it. In sparse reward setting, we further modify PEARL with Hindsight Experience Replay [1] for a fair comparison[2]. The last one means we train a shared HAC policy jointly across all meta-train tasks sampled from the whole task distribution. In addition, for a fair comparison, we modify the HAC source code with SAC algorithm which is considered to be much powerful than DDPG in the original implementation [11], to ensure the consistency to PEARL and MGHRL.

We set the goal space to be the set of all possible positions of the gripper, in which a goal is a 3-d vector. In the environments, the low level policy of our algorithm aims to move the gripper to the desired goal position. Such policy won't change at all when switching to other tasks since the mechanism of moving gripper keeps the same between different tasks. Thus we use a shared policy trained jointly across all tasks for the low level of MGHRL. We set the maximum low-level horizon K to be 10 and the distance threshold to be 0.05. The high level context data sampler S_c^h samples uniformly from the most recently collected batch of data, which is recollected every 1000 meta-training steps. Unlike HAC, we use target networks for both levels, which updates with $\tau = 0.005$. All context encoder, actor and critic neural networks had three hidden layers, with 300 nodes in each layer. The discount factor was set to $\gamma = 0.99$. We use a sparse reward function in which a reward of 0 is granted only if the terminal goal given by the environment is achieved and a reward of -1 otherwise. The dense reward used in our baseline is a value corresponding to the distance between current position of the box (or gripper in fetch-reach case) and desired goal position. We do our experiments on 50 train tasks and 10 test tasks, where the difference between each task is in the terminal goal position we want the box or gripper to reach.

5.2   Results

We first do the simplest meta-learning evaluation on each type of the four tasks. In each scenario, we evaluate on 50 meta-train tasks and 10 meta-test tasks, where the difference between each task is in the terminal goal position we want the box or gripper to reach as well as the initial positions. The results are shown in Table 1. In Fetch-reach environment which is very easy to learn as we mentioned before, the tested methods all reach a final performance of 100% success rate. Our method MGHRL outperforms the other three methods in Push and Slide scenarios, while PEARL with dense reward performs better in Pick-Place tasks. Our two-level hierarchy and hindsight transitions significantly decrease the difficulty of meta learning with sparse reward, and is able to learn efficiently under a fixed budget of environment interactions. HAC with shared policy lacks

---

[2] We also evaluated PEARL (without HER) with sparse reward and it was not able to solve any of the tasks.

generalization ability and cannot always achieve good performance when tested on varied tasks as shown in our results. We assume that it is because in our settings since we only change the terminal goals' positions to create different tasks, thus it is possible that the policy learned from one task will work on other task whose terminal goal positions are very close to previous training ones. But such method lacks generalization ability and cannot always achieve good performance when tested on varied tasks as shown in our results.

Table 1. Average success rates over all tasks (meta-test)

Tasks	MGHRL	PEARL	HER-PEARL	HAC
Reach	100%	100%	100%	100%
Push	**76%**	61%	15%	41%
Slide	**36%**	5%	6%	23%
Pick-Place	92%	**98%**	47%	13%

We further evaluate our method on tasks with wider distribution. As shown in Fig. 3, each scenario's meta-train and meta-test tasks are sampled from the original two or three types of tasks (e.g. 30 meta-train tasks from Push and 30 meta-train tasks from Slide). Our algorithm MGHRL generally achieves better performance and adapts to new task much more quickly in all four types of scenarios. Directly using PEARL to learn a meta-policy that considers both overall strategy and detailed execution mechanism would decrease prediction accuracy and sample efficiency in these wider-distribution tasks as shown empirically. It is better to decompose the meta training process and focus on goal-generation learning. In this way, our agent only needs to learn a meta-policy that gives the learning rules for learning how to generate proper subgoals. Moreover, under dense reward settings of these challenging tasks, the critic of PEARL has to approximate a highly non-linear function that includes the Euclidean distance between positions and the difference between two quaternions for rotations [19]. Using the spars e return is much simpler since the critic only has to differentiate between successful and failed states.

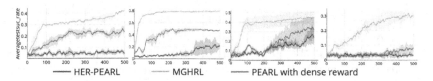

Fig. 3. Average success rates for MGHRL, PEARL agents in each scenario: (a) Push & Slide & Pick-Place, (b) Push & Pick-Place, (c) Pick-Place & Slide, (d) Push & Slide. Each algorithm was trained for 1e6 steps. The error bar shows 1 standard deviation.

6   Discussion and Future Work

In this paper, we propose a hierarchical meta-RL algorithm, MGHRL, which realizes meta goal-generation and leaves the low-level policy for independent RL. MGHRL focuses on learning the overall strategy of tasks instead of learning detailed action execution mechanism to improve the efficiency and generality. Our experiments show that MGHRL outperforms the SOTA especially in problems with relatively wider task distribution. Beyond this paper, we believe our algorithm can accelerate the acquisition of entirely new tasks. For example, to learn tasks such as riding bicycles and riding a motorcycle, the two primitive action execution mechanisms are entirely different but the two learning processes still share similar high-level structures (e.g. how to keep balance). With meta learning on high level policy, our algorithm is still supposed to achieve good performance on such tasks. We leave these for future work to explore.

References

1. Andrychowicz, M., et al.: Hindsight experience replay. In: Advances in Neural Information Processing Systems 30: Annual Conference on Neural Information Processing Systems 2017, pp. 5048–5058 (2017). http://papers.nips.cc/paper/7090-hindsight-experience-replay
2. Bacon, P., Harb, J., Precup, D.: The option-critic architecture. In: Proceedings of the Thirty-First AAAI Conference on Artificial Intelligence, pp. 1726–1734 (2017). http://aaai.org/ocs/index.php/AAAI/AAAI17/paper/view/14858
3. Barto, A.G., Mahadevan, S.: Recent advances in hierarchical reinforcement learning. Discrete Event Dyn. Syst. **13**(1–2), 41–77 (2003). https://doi.org/10.1023/A:1022140919877
4. Bengio, Y., Bengio, S., Cloutier, J.: Learning a synaptic learning rule. In: IJCNN-91-Seattle International Joint Conference on Neural Networks II, vol. 2, p. 969 (1991)
5. Bengio, Y., LeCun, Y. (eds.): 4th International Conference on Learning Representations, ICLR 2016 (2016). https://iclr.cc/archive/www/doku.php%3Fid=iclr2016:accepted-main.html
6. Dayan, P., Hinton, G.E.: Feudal reinforcement learning. In: Advances in Neural Information Processing Systems 5, [NIPS Conference], pp. 271–278 (1992). http://papers.nips.cc/paper/714-feudal-reinforcement-learning
7. Dietterich, T.G.: Hierarchical reinforcement learning with the MAXQ value function decomposition. J. Artif. Intell. Res. **13**, 227–303 (2000). https://doi.org/10.1613/jair.639
8. Duan, Y., Schulman, J., Chen, X., Bartlett, P.L., Sutskever, I., Abbeel, P.: RL^2: fast reinforcement learning via slow reinforcement learning. CoRR abs/1611.02779 (2016). http://arxiv.org/abs/1611.02779
9. Finn, C., Abbeel, P., Levine, S.: Model-agnostic meta-learning for fast adaptation of deep networks. In: Proceedings of the 34th International Conference on Machine Learning, ICML 2017, pp. 1126–1135 (2017). http://proceedings.mlr.press/v70/finn17a.html
10. Frans, K., Ho, J., Chen, X., Abbeel, P., Schulman, J.: Meta learning shared hierarchies. In: 6th International Conference on Learning Representations, ICLR 2018 (2018)

11. Haarnoja, T., Zhou, A., Abbeel, P., Levine, S.: Soft actor-critic: off-policy maximum entropy deep reinforcement learning with a stochastic actor. In: Proceedings of the 35th International Conference on Machine Learning, ICML 2018, pp. 1856–1865 (2018). http://proceedings.mlr.press/v80/haarnoja18b.html
12. Kingma, D.P., Welling, M.: Auto-encoding variational bayes. In: 2nd International Conference on Learning Representations, ICLR 2014 (2014). http://arxiv.org/abs/1312.6114
13. Levine, S., Finn, C., Darrell, T., Abbeel, P.: End-to-end training of deep visuomotor policies. J. Mach. Learn. Res. **17**, 39:1–39:40 (2016). http://jmlr.org/papers/v17/15-522.html
14. Levy, A., Konidaris, G., Platt Jr, R., Saenko, K.: Learning multi-level hierarchies with hindsight. In: 7th International Conference on Learning Representations, ICLR 2019 (2019). https://openreview.net/forum?id=ryzECoAcY7
15. Mishra, N., Rohaninejad, M., Chen, X., Abbeel, P.: A simple neural attentive meta-learner. In: 6th International Conference on Learning Representations, ICLR 2018 (2018). https://openreview.net/forum?id=B1DmUzWAW
16. Mnih, V., et al.: Human-level control through deep reinforcement learning. Nature **518**(7540), 529–533 (2015). https://doi.org/10.1038/nature14236
17. Nachum, O., Gu, S., Lee, H., Levine, S.: Data-efficient hierarchical reinforcement learning. In: Advances in Neural Information Processing Systems 31: Annual Conference on Neural Information Processing Systems 2018, NeurIPS 2018, pp. 3307–3317 (2018). http://papers.nips.cc/paper/7591-data-efficient-hierarchical-reinforcement-learning
18. Parr, R., Russell, S.J.: Reinforcement learning with hierarchies of machines. In: Advances in Neural Information Processing Systems 10, [NIPS Conference], pp. 1043–1049 (1997). http://papers.nips.cc/paper/1384-reinforcement-learning-with-hierarchies-of-machines
19. Plappert, M., et al.: Multi-goal reinforcement learning: challenging robotics environments and request for research. CoRR abs/1802.09464 (2018). http://arxiv.org/abs/1802.09464
20. Rakelly, K., Zhou, A., Finn, C., Levine, S., Quillen, D.: Efficient off-policy meta-reinforcement learning via probabilistic context variables. In: Proceedings of the 36th International Conference on Machine Learning, ICML 2019, pp. 5331–5340 (2019). http://proceedings.mlr.press/v97/rakelly19a.html
21. Ravi, S., Larochelle, H.: Optimization as a model for few-shot learning. In: ICLR (2017)
22. Rothfuss, J., Lee, D., Clavera, I., Asfour, T., Abbeel, P.: ProMP: proximal meta-policy search. In: 7th International Conference on Learning Representations, ICLR 2019 (2019). https://openreview.net/forum?id=SkxXCi0qFX
23. Santoro, A., Bartunov, S., Botvinick, M., Wierstra, D., Lillicrap, T.P.: Meta-learning with memory-augmented neural networks. In: Proceedings of the 33nd International Conference on Machine Learning, ICML 2016, pp. 1842–1850 (2016). http://proceedings.mlr.press/v48/santoro16.html
24. Schmidhuber, J.: Evolutionary principles in self-referential learning (1987)
25. Stadie, B.C., et al.: Some considerations on learning to explore via meta-reinforcement learning. CoRR abs/1803.01118 (2018). http://arxiv.org/abs/1803.01118
26. Sutton, R.S., Precup, D., Singh, S.P.: Between MDPs and semi-MDPs: a framework for temporal abstraction in reinforcement learning. Artif. Intell. **112**(1–2), 181–211 (1999). https://doi.org/10.1016/S0004-3702(99)00052-1

27. Thrun, S., Pratt, L.Y.: Learning to Learn. Springer, Boston (1998). https://doi. org/10.1007/978-1-4615-5529-2
28. Todorov, E., Erez, T., Tassa, Y.: MuJoCo: a physics engine for model-based control. In: 2012 IEEE/RSJ International Conference on Intelligent Robots and Systems, IROS 2012, pp. 5026–5033 (2012). https://doi.org/10.1109/IROS.2012.6386109
29. Vezhnevets, A.S., et al.: Feudal networks for hierarchical reinforcement learning. In: Proceedings of the 34th International Conference on Machine Learning, ICML 2017, pp. 3540–3549 (2017). http://proceedings.mlr.press/v70/vezhnevets17a.html
30. Vinyals, O., Blundell, C., Lillicrap, T., Kavukcuoglu, K., Wierstra, D.: Matching networks for one shot learning. In: Advances in Neural Information Processing Systems 29: Annual Conference on Neural Information Processing Systems 2016, pp. 3630–3638 (2016). http://papers.nips.cc/paper/6385-matching-networks-for-one-shot-learning
31. Wang, J.X., et al.: Learning to reinforcement learn. CoRR abs/1611.05763 (2016). http://arxiv.org/abs/1611.05763
32. Xu, T., Liu, Q., Zhao, L., Peng, J.: Learning to explore via meta-policy gradient. In: Proceedings of the 35th International Conference on Machine Learning, ICML 2018, pp. 5459–5468 (2018). http://proceedings.mlr.press/v80/xu18d.html

D3PG: Decomposed Deep Deterministic Policy Gradient for Continuous Control

Yinzhao Dong[1] ⓘ, Chao Yu[2](✉) ⓘ, and Hongwei Ge[1] ⓘ

[1] Dalian University of Technology, Dalian 116024, China
[2] Sun Yat-sen University, Guangzhou 510275, China
yuchao3@mail.sysu.edu.cn

Abstract. In this paper, we study how structural decomposition and multiagent interactions can be utilized by deep reinforcement learning in order to address high dimensional robotic control problems. In this regard, we propose the D3PG approach, which is a multiagent extension of DDPG by decomposing the global critic into a weighted sum of local critics. Each of these critics is modeled as an individual learning agent that governs the decision making of a particular joint of a robot. We then propose a method to learn the weights during learning in order to capture different levels of dependencies among the agents. The experimental evaluation demonstrates that D3PG can achieve competitive or significantly improved performance compared to some widely used deep reinforcement learning algorithms. Another advantage of D3PG is that it is able to provide explicit interpretations of the final learned policy as well as the underlying dependencies among the joints of a learning robot.

Keywords: Deep reinforcement learning · Structural decomposition · Interpretability · Robotic control

1 Introduction

The integration of deep neural networks into reinforcement learning (RL) [32] has fostered a new flourishing research area of *Deep RL* (DRL) [17]. A plethora of new algorithms, such as DDPG [18], TRPO [28], A3C [23] and GAE [29], have been proposed in recent years to address a wide range of challenging continuous control problems in robotic locomotion and manipulation. However, since the existing DRL algorithms directly search in the entire high dimensional continuous state/action space and output the learned policy in an end-to-end manner, the learning efficiency is still quite low. In addition, it is usually difficult to provide meaningful interpretations on the performance of these learning algorithms due to lack of deeper investigations of the structures of the learned policies.

In this paper, we resort to a decomposition scheme for more efficient and interpretable learning in robotic continuous control problems. This decomposition

Supported by National Natural Science Foundation of China under Grant No. 62076259.

M. E. Taylor et al. (Eds.): DAI 2020, LNAI 12547, pp. 40–54, 2020.
https://doi.org/10.1007/978-3-030-64096-5_4

scheme is motivated by the fact that real-world robots, such as robotic manipulators and multi-legged robots, can be structurally decomposed into a number of subcomponents. These subcomponents are able to manage their resources independently, while working together to make the robot perform a desired task [5, 16]. Thus, the decomposition scheme enables automatically decomposing the complex learning problem into local, more readily learnable sub-problems so that the original learning problem can be greatly simplified. For example, a robotic arm can be structurally decomposed into three subcomponents of upper arm, forearm and palm, which are connected by the joints of shoulder, elbow and wrist, respectively. The whole decision making problem then can be decomposed into several sub-problems, each of which corresponds to controlling one particular variable of setting the angle values for each of these three joints. In this regard, if the decision making for each sub-problem is controlled by an individual learning agent, then it is able to apply Multi-Agent Learning (MAL) [1] methods to coordinate these agents' behaviors for a maximum global performance for the whole robot.

In this work, we propose a novel algorithm, *Decomposed Deep Deterministic Policy Gradient* (D3PG), for decomposed learning in continuous robotic control. In D3PG, the global critic is decomposed into a weighted sum of some local critics, each of which is modeled as an individual agent that depends only on its own observations and actions. Each critic is updated by backpropagating gradients using the joint global reward, i.e., the critic is learned implicitly rather than from any reward specific to the agent, thus we do not impose constraints that the critics are action-value functions for any specific reward. The weights in the summation measure the relative importance of the critics during the learning process. Thus, dynamical adaptation of these weights implicitly addresses the credit assignment puzzle in general MAL problems. In order to compute the weights, we first use a *coordination graph* (CG) to synchronize the interactions among the agents. Each edge on this graph indicates that the two linked agents need to coordinate over their behaviors for better learning performance, and the weight on each edge indicates the extent of such a coordination. In this regard, we then propose a *Prediction CG* (PCG) method to compute the weights of CG during learning using the prediction errors in other agents' states and build a graph that links those agents with the largest weights.

Experimental evaluations in typical Mujoco environments [33] verify the effectiveness of the proposed D3PG algorithm in various robotic control problems. In addition, by analyzing the weights of the CG after convergence, explicit interpretations can be obtained regarding the final learned policy as well as the underlying roles of the components of a robot in different motion postures.

2 Background

2.1 Reinforcement Learning (RL)

An RL problem [32] is typically modeled as a Markov Decision Process (MDP), which can be defined by the tuple (S, A, T, R), where S and A are respectively the set of agent's states and actions, $T(s, a, s') = P(s'|s, a)$ gives the probability of jumping to the new state s' given current state s and action a, and $R(s, a)$ is the reward function. An agent's behavior is defined by a policy $\pi(s, a)$ which maps states to a probability distribution over the actions. The aim of RL is to find the best policy $\pi^*(s, a)$ to maximize the cumulative discount reward $G = \sum_t \gamma^t r_t$, where r_t is the reward at time t, and $0 < \gamma < 1$ is a discount factor. Given an MDP and an agent's policy, we then define the action-value function $Q^\pi(s, a)$, which denotes the cumulative discount reward after taking action a in state s and then using policy π to explore. Deep RL (DRL) [17] adopts a deep neural network to approximate the value functions, combined with some enabling techniques such as experience replay and target network to stabilize learning and improve data efficiency.

2.2 Deep Deterministic Policy Gradient (DDPG)

DDPG algorithm [18] is a classic actor-critic DRL algorithm that enables learning in continuous domains. There is a main network consisting of a critic network $Q(s, a|\theta^Q)$ and an actor network $\mu(s|\theta^\mu)$. The parameter θ^Q is optimized to minimize the loss L:

$$L = E_{s_i, a_i, r_i, s_{i+1} \sim D}(y_i - Q(s_i, a_i|\theta^Q))^2 \tag{1}$$

where $y_i = r_i + \gamma Q'(s_{i+1}, \mu'(s_{i+1}|\theta^{\mu'})|\theta^{Q'})$ is a target value computed by separate target networks of $Q'(s, a|\theta^{Q'})$ and $\mu'(s|\theta^{\mu'})$, and D is an experience replay buffer. The parameters θ^μ are optimized with the gradient given as follows:

$$\nabla_{\theta^\mu} J(\theta) = E_{s \sim D}[\nabla_{\theta^\mu} \mu(s|\theta^\mu) \nabla_a Q(s, a_\mu|\theta^Q)] \tag{2}$$

where $a_\mu = \mu(s|\theta^\mu)$ is the action computed by the actor.

The policy network outputs an action deterministically. In order to enable exploration, DDPG constructs a behavior policy by adding noise sampled from a noisy process $\mathcal{N}$ to the actor policy as $\beta = \mu(s|\theta^\mu) + \mathcal{N}$.

3   The D3PG Algorithm for Robotic Control

In this section, we propose a multiagent extension of DDPG for efficient learning in robotic control problems. The novel algorithm, D3PG, enables a decomposition of the global critic value function into a set of local critic value functions according to the inherent robot structures. In D3PG, each local critic is modeled as an individual learning agent that depends only on its own observations and actions. Thus, the original learning problem in the whole search space can be greatly simplified by searching in some additive local spaces. We first show how to structurally decompose a robot and compute the relative importance of each decomposed component, and then provide a weight computation method and the training procedure of D3PG.

Fig. 1. The illustration of the structural decomposition in two robot environments: Walker2d (left) and Half-cheetah (right).

3.1 Structural Decomposition

In the real world, a robot can be naturally decomposed into multiple sub-components (agents) that are connected by their in-between joints according to its physical structure. These agents are able to manage their resources independently, while working together to make the robot perform a desired task. To simplify notation, the whole robot's state $s \in \mathbb{R}^M$ and action $a \in \mathbb{R}^N$ can be decomposed as follows:

$$
\begin{aligned}
s^1, \ldots s^l, \ldots s^n, s^g &= Divide(s) \\
a^1, \ldots a^l, \ldots a^n &= Divide(a)
\end{aligned}
\tag{3}
$$

where $Divide$ is a decomposition operator, s^g denotes a set of state information that is shared by all the agents, and n denotes the number of agents, while $s^l \in \mathbb{R}^{M^l}$ and $a^l \in \mathbb{R}^{N^l}$ denote the state and action of the l-th agent, respectively.

Figure 1 illustrates the structural decomposition in two robot environments. For example, Walker2d can be divided into six agents, each of which represents a key joint. Among the 17-dimensional states and 6-dimensional actions, 5 global states (2 for position states and 3 for velocity states) are shared by all the agents. Each agent then has a 2-dimensional local state space indicating the position and speed of a joint and 1-dimensional action controlling the angle of this joint, thus significantly reducing the overall complexity compared to directly searching in the whole state/action space.

3.2 The PCG Method

The local agents after decomposition play varying roles in the control of the whole robot. For example, during walking, the robot is simultaneously controlled by the torso, thighs and calves, as well as the knee joints and ankle joints. The ankle agents should make decisions based on the information of other joint agents in order to maintain balance. Therefore, to guarantee a smooth motion, the key joints must coordinate with each other over setting their values. To enable such coordination, we use a *coordination graph* (CG) to model the dependencies among different agents, which is formally given by $G = (\mathcal{V}, \mathcal{W})$, where $\mathcal{V} = \{A_l | l \in [1, n]\}$ is a set of agents with A_l denoting the l-th agent and $\mathcal{W}$ denotes a weight matrix with each element w_{lj} indicating the relative importance of agent

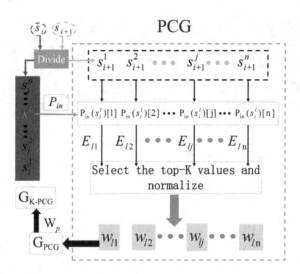

Fig. 2. The PCG method, where s_i^l denotes the state of the l-th agent in the i-th step.

A_j considered by agent A_l ($\sum_{j=1}^n w_{lj} = 1$). Once the CG can be obtained, the overall weight of each agent A_l can be computed as $W^l = \sum_{j=1}^n w_{jl}$. We then introduce the PCG method in Fig. 2 to compute the weights of the agents during the learning process.

The PCG method computes CG by a predictor P_{in} (e.g., a neural network) with an input of the state of the current agent A_l. Each agent predicts the state of other agents in the next step based on its own state, and utilizes the difference between the predicted states and the real states to represent the weight. The predicted errors E_{lj} can be given as follows:

$$E_{lj} = \left\| s_{i+1}^j - P_{in}(s_i^l)[j] \right\|_2 \tag{4}$$

where s_{i+1}^j and $P_{in}(s_i^l)[j]$ denote the real state and the predicted state of agent A_j at step $i + 1$, respectively, and s_i^l denotes the state vector of A_l at step i.

A higher E_{lj} means that the behaviors of A_j are more unpredictable and should be paid more attention by agent A_l in the next step. Thus, E_{lj} can represent the weight w_{lj} between these two agents. In order to reduce the structure complexity of CG, we then select the top-K highest E_{lj} values to build a graph G_{K-PCG}, where K is a hyper-parameter controls the number of edges. Let the adjacency matrix of G_{K-PCG} be $\mathcal{W}_{K-PCG}$. The final $\mathcal{W}_{PCG}$ can be obtained using the following update rule:

$$\mathcal{W}_{PCG} = \eta * \mathcal{W}_{K-PCG} + (1 - \eta) * \mathcal{W}_P. \tag{5}$$

where η is a parameter to control the bias of selection, and $\mathcal{W}_p$ denotes the adjacency weight matrix based on the physical structure of the robot, which means that all the physically linked agents have a fixed equal weight to a local agent.

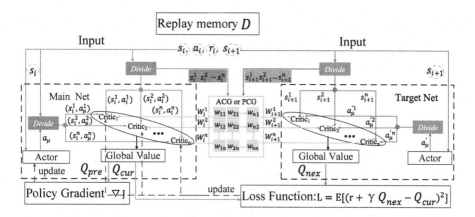

Fig. 3. The flowchart of the D3PG algorithm.

During the training process, the weights ρ of the state predictor P_{in} can be optimized through the loss L_ρ:

$$L_\rho = \sum_{j=1}^{n} \sum_{l=1}^{n} (s_{i+1}^j - P_{in}(s_i^l)[j])^2. \tag{6}$$

The PCG method can reduce the computational complexity by building a much sparser graph and thus capturing the most important dependencies among the agents. Moreover, it enables a wider exploration of the weights over the original structure, as given by Eq. (5).

3.3   The D3PG Algorithm

We then introduce the decomposed multiagent extension of the DDPG algorithm. Figure 3 plots the flowchart of the algorithm. At each time step i, the algorithm samples a $Minibatch$ size of transitions s_i, a_i, r_i, s_{i+1} from the replay buffer D. Then the s_i, a_i and s_{i+1} are divided according to Eq. (3). In the Main Net, the Actor takes the state s_i as input and outputs the joint action $a_\mu = \mu(s_i|\theta^\mu)$ of the whole robot. The global Critic is decomposed into n critic networks $Q_l(s^l, a^l)$, each of which is conditioned on the local state s^l and action a^l of each agent A_l:

$$Q = \sum_{l=1}^{n} (W^l * Q_l(s^l, a^l|\theta^{Q_l})). \tag{7}$$

The Actor aims to learn a policy with regard to parameters θ^μ that maximize the expected discounted return J:

$$\nabla_{\theta^\mu} J = \nabla_{\theta^\mu} \mu(s_i|\theta^\mu) \nabla_{a_\mu} Q_{pre} \tag{8}$$

$$Q_{pre} = \sum_{l=1}^{n} (W_i^l * Q_l(s_i^l, a_\mu^l|\theta^{Q_l})) \tag{9}$$

where Q_{pre} denotes the global value based on the current state s_i and the joint action a_μ.

Each critic network $Q_l(s_i, a_i | \theta^{Q_l})$ is then updated to optimize the parameter θ^{Q_l} by minimizing the following loss:

$$L = (r_i + \gamma Q_{nex} - Q_{cur})^2 \tag{10}$$

$$Q_{cur} = \sum_{l=1}^{n} (W_i^l * Q_l(s_i^l, a_i^l | \theta^{Q_l})) \tag{11}$$

$$Q_{nex} = \sum_{l=1}^{n} (W_{i+1}^l * Q_l'(s_{i+1}^l, a_\mu'^l | \theta^{Q_l}) \tag{12}$$

where Q_{cur} is the global value conditioned on the current state s_i and current action a_i, Q_{nex} is the global value based on the next state s_{i+1} and next joint action $a_\mu' = \mu'(s_{i+1} | \theta^{\mu'})$ computed by the target Actor, r_i denotes the current reward, and γ denotes the reward decay.

The weights of the Target Net are updated to slowly track the learned networks as follows:

$$\theta^{Q_l'} \leftarrow \tau \theta^{Q_l} + (1 - \tau)\theta^{Q_l'}, l \in [0, n] \tag{13}$$

$$\theta^{\mu'} \leftarrow \tau \theta^{\mu} + (1 - \tau)\theta^{\mu'} \tag{14}$$

The complete pseudo-code of the D3PG algorithm is given in Algorithm 1.

4   Experiment

The MuJoCo [33] platform in OpenAI gym is used to conduct our experiments. We compare D3PG to some of the most famous DRL algorithms in the literature [4], including PPO [30], DDPG [18], AC [25] and REINFORCE [37]. Each simulation run lasts 1500 episodes and an episode lasts for 300 steps or until the robot falls down. The final results are averaged over 5 independent runs. The main parameters regarding the algorithms are tuned for optimal performance and are listed in Table 1. Parameter η is set the same value as γ. The settings of the neural networks used in D3PG are given in Table 2, while Table 3 shows the state and action dimensions in the four robot environments.

Figure 4 shows the learning dynamics using D3PG compared to different DRL algorithms, where D3PG-ACG is the method that learns the CG using the attention mechanism [35], and the parameter K in D3PG-PCG is also tuned to generate the optimal performance. It is clear that D3PG-PCG can achieve a faster convergence rate than the other algorithms, except in Swimmer where PPO dominates. This result is mainly due to the relatively small size of Swimmer. As the environment size gets larger, the performance gap between D3PG-PCG and the DRL algorithms becomes more apparent, while the performance of D3PG-ACG decreases due to the high complexity in computing a full graph. Table 4 further compares the rewards in the beginning, intermediate and final

Algorithm 1. The D3PG Algorithm

Input: Parameter: Initialize n critics $Q_l(s, a|\theta^{Q_l}), l \in [0, n]$, and Actor $\mu(s|\theta^\mu)$ with weights θ^{Q_l} and θ^μ, receptively. Initialize Target Net with Q'_l, and μ' with weights $\theta^{Q'_l}$ and $\theta^{\mu'}$, receptively. Initialize State Predictor with ρ, and replay buffer D.

1: **for** $ep = 1, epsiode$ **do**
2:      Initialize a random process $\mathcal{N}$ for choosing actions;
3:      Receive initial observation state s_0;
4:      **for** $t = 1, T$ **do**
5:          Select action $a_t = \mu(s_t|\theta^\mu) + \mathcal{N}$;
6:          Obtain reward r_t and new state s_{t+1};
7:          Store transition (s_t, a_t, r_t, s_{t+1}) in D
8:      **end for**
9:      **for** $it = 1, Iteration$ **do**
10:          Randomly sample a $Minibatch$ size of transitions (s_i, a_i, r_i, s_{i+1}) from D;
11:          Execute PCG and compute W_i and W_{i+1};
12:          $Divide$ s_i, s_{i+1}, a_i, a_μ, and a'_μ, respectively;
13:          $Q_{cur} = \sum_{l=1}^n (W_i^l * Q_l(s_i^l, a_i^l|\theta^{Q_l}))$;
14:          $Q_{pre} = \sum_{l=1}^n (W_i^l * Q_l(s_i^l, a_\mu^l|\theta^{Q_l}))$;
15:          $Q_{nex} = \sum_{l=1}^n (W_{i+1}^l * Q'_l(s_{i+1}^l, a_\mu'^l|\theta^{Q'_l})$;
16:          Update the n critics by the loss:
17:              $L = (r_i + \gamma Q_{nex} - Q_{cur})^2$
18:          Update the actor by the sampled policy gradient:
19:              $\nabla_{\theta^\mu} J = \nabla_{\theta^\mu} \mu(s_i|\theta^\mu) \nabla_{a_\mu} Q_{pre}$;
20:          Update the target networks:
21:              $\theta^{Q'_l} \leftarrow \tau\theta^{Q_l} + (1 - \tau)\theta^{Q'_l}, l \in [0, n]$;
22:              $\theta^{\mu'} \leftarrow \tau\theta^\mu + (1 - \tau)\theta^{\mu'}$;
23:          Update the State Predictor by the loss:
24:              $L_\rho = \sum_{j=1}^n \sum_{l=1}^n (s_{i+1}^j - P_{in}(s_i^l)[j])^2$
25:      **end for**
26: **end for**

---

period of the learning process. We can discover that D3PG-PCG usually does not have advantages in the early period of learning as it takes some time for the agents to learn the weight values. As the learning proceeds, D3PG-PCG then outperforms the other algorithms in the later period of learning process. The minor difference between D3PG-PCG and DDPG in Hopper at step 1500 is due to the larger fluctuation of D3PG-PCG during the learning process. However, the superior of D3PG-PCG can be observed from the learning curves in Fig. 4(b)(f).

Figure 5 shows the learning performance of and D3PG-ACG, D3PG-PCG (with different K values) and D3PG-CG (which uses the fixed physical structure G_P as the CG). Since D3PG-ACG uses the fixed physical structure G_P as the optimization goal, it has similar performance with D3PG-CG. Both algorithms perform worse than D3PG-PCG and the performance gap gets larger when the environment becomes more complex. The physical structure cannot capture the varying of weights during different postures in the learning process. This is contradictory to D3PG-PCG, which enables a wider exploration over the

Table 1. The main parameters in different RL algorithms.

Parameters	AC	DDPG	REINFORCE	PPO	D3PG
γ	0.99	0.99	0.99	0.99	0.99
λ	0.001	0.001	0.001	0.001	0.001
$Hiddenlayer$	32	400	20	64	400
$Replaysize$	None	$5e^4$	None	$1e^4$	$5e^4$
$Minibatch$	None	64	None	32	1
$Iteration$	1	10	1	5	10
$Clip$	None	None	None	0.2	None

Table 2. The settings of neural networks in D3PG.

Neural network	Input	First layer	Second layer	Output
Critic network	State dimensions	64	None	1
Actor network	State dimensions	400	300	1
State predictor	State dimensions	128	None	n*State dimensions
Feature MLP	State dimensions	None	None	36 (b)
Attention MLP	36	None	None	1

physical structure, such that the implicit influence of other unlinked agents can be considered.

Finally, Fig. 6 shows the weights of the joints in Walker2d when it starts moving and keeps running. At the beginning of moving, the weights of the two feet (A_3 and A_6) play the most important and stable roles, while the weights of other joints change dynamically in order to keep the robot walk stably. When the robot keeps running, the two feet are still the most important parts, followed by the thighs (A_1 and A_4) and then the knees (A_2 and A_5). It can be interpreted that the weights tend to be stable when the robot is conducting a smooth moving behavior such as running, but changing dynamically during a slow motion that requires more careful coordination among different agents in different postures.

5   Related Work

Some multiagent DRL methods have been developed in recent years under the setting of *centralised learning with fully decentralised execution*. The COMA algorithm [7] utilizes a centralized critic to train decentralised actors and estimate a counterfactual advantage function for each agent to reduce the variance of policy gradient. Gupta et al. [9] present a centralised actor-critic algorithm with a critic for each agent. Lowe et al. [19] proposed the MADDPG algorithm which learns a centralised critic for each agent and applies this to competitive games with continuous action spaces. Unlike these approaches where a centralized critic is

Table 3. The state and action dimensions in the robot environments.

Robot environments	State dimensions	Action dimensions
Swimmer	8	2
Hopper	11	3
Walker	17	6
Half-Cheetah	17	6

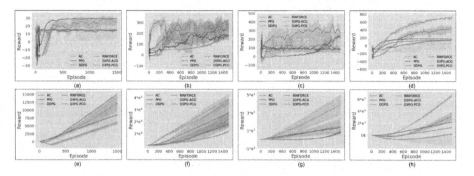

Fig. 4. The comparison to some DRL algorithms in Swimmer, Hopper, Walker and Half-Cheetah, respectively, in terms of average reward (the upper row) and accumulated reward (the lower row).

maintained by each agent or shared by all the agents, D3PG decomposes the centralized critic into some local critics that are based only on the local information of each agent.

Value decomposition has been widely applied in MAL research. Particularly, studies have focus on cooperative learning on CG by decomposing the global value functions into some local ones in order to reduce the computation complexity [8,13]. Several studies addressed decomposed learning in multiagent DRL [26,31]. Unlike these studies, D3PG learns the value decomposition autonomously from experience, and addresses continuous control problems.

Decomposing a single robot learning problem into an MAL problem has been studied before. Busoniu *et al.* [2] investigated centralized and decentralized MAL in solving a two-link manipulator control problem. Additionally, Martin and De Lope [21] presented a distributed MAL modeling for generating a real-time trajectory of a robot. The authors [5,34] applied independent learning methods for robotic control problems. Kabysh *et al.* [12] proposed an MAL approach to control a robotic arm. Similarly, Matignon *et al.* [22] proposed a semi-decentralized MAL approach for controlling a robotic manipulator. Leottau *et al.* [14,15] applied concurrent layered learning strategies in the context of soccer robotics and proposed a five-stage methodology to solve a robotic learning control problem. Yu *et al.* [38] proposed a multiagent extension of NAF in robotic control. Unlike all the existing approaches, where agents either learn indepen-

Table 4. The rewards in different periods of learning process using different algorithms (the highest values are marked in bold).

MuJoCo	Episodes	AC	PPO	DDPG	REINFORCE	D3PG-ACG	D3PG-PCG
Swimmer	1	15.2 ± 2.6	-8.6 ± 1.8	8.8 ± 0.9	$\mathbf{15.5 \pm 3.2}$	6.2 ± 2.8	-6.2 ± 5.5
	750	14.2 ± 0.2	$\mathbf{28.7 \pm 0.2}$	15.1 ± 2.0	14.2 ± 0.1	20.5 ± 22.8	15.5 ± 3.2
	1500	14.7 ± 0.8	$\mathbf{28.4 \pm 0.5}$	13.8 ± 1.1	14.7 ± 0.9	19.1 ± 24.8	25.2 ± 12.6
Hopper	1	42.1 ± 1.0	36.3 ± 0	-0.1 ± 0.5	$\mathbf{45.3 \pm 1.6}$	22.5 ± 0.1	13.0 ± 0.1
	750	58.8 ± 2.5	168.0 ± 0	107.3 ± 117.0	207.0 ± 168.6	221.0 ± 176.7	$\mathbf{294.2 \pm 117.9}$
	1500	189.6 ± 4.3	175.6 ± 0	$\mathbf{285.4 \pm 51.1}$	232.6 ± 193.8	242.3 ± 220.5	249.4 ± 172.7
Walker2d	1	$\mathbf{69.8 \pm 12.4}$	-7.9 ± 0.5	-0.1 ± 0.2	67.2 ± 17.4	0.6 ± 5.2	1.6 ± 0.2
	750	92.1 ± 4.7	22.8 ± 1.85	122.5 ± 133.3	99.5 ± 16.8	151.2 ± 72.6	$\mathbf{290.0 \pm 182.4}$
	1500	82.3 ± 3.0	234.5 ± 62.9	82.1 ± 93.1	5.5 ± 85.8	139.2 ± 76.9	$\mathbf{281.9 \pm 134.4}$
Half-cheetah	1	-234.4 ± 7.8	-96.9 ± 5.8	$\mathbf{-0.9 \pm 0.9}$	-292.8 ± 2.7	-3.7 ± 0.8	-7.8 ± 1.0
	750	117.9 ± 20.8	171.3 ± 12.4	274.3 ± 136.0	46.5 ± 23.5	-55.7 ± 114.9	$\mathbf{727.5 \pm 271.8}$
	1500	154.0 ± 9.3	179.3 ± 2.3	423.9 ± 216.0	146.9 ± 5.8	76.7 ± 241.5	$\mathbf{894.8 \pm 213.6}$

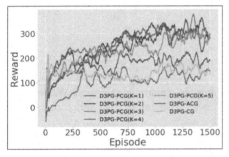 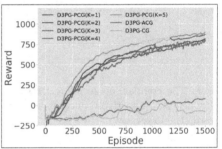

Fig. 5. The learning performance of different D3PG algorithms in Walker2d (left) and Half-Cheetah (right).

dently without any explicit coordination, or coordinate with each other in a fixed procedure (e.g., by sharing a fixed amount of information), D3PG models the dynamic dependencies among the agents through computing the continuously changing weights of each local critic. Several studies also applied *Graph Neural Networks* (GNNs) to learn structured policies for DRL in robotic control [27,36]. Again, they did not model the varying importance of different components in the robot.

Some other studies have investigated in communication learning in the MAL literature [6,11,24]. However, their focus is mainly on learning when communication is needed in multiagent systems. Moreover, several studies also applied attention mechanisms in MAL. The main idea is to learn a centralized critic over other agents' states and actions with an attention mechanism in the actor-critic algorithm [10,20], or use attention mechanisms to generate enhanced state embedding in PPO algorithm [3]. In D3PG, the attention mechanism is used to compute the weights of the local critics, in order to address the credit assignment problem in global critic learning.

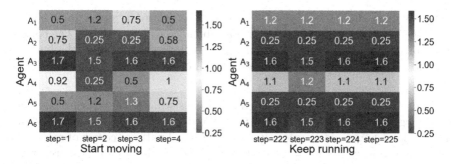

Fig. 6. The importance of each agent (joint) in Walker2d at the start of moving (the left) and during running (the right).

6   Conclusions

In this paper, we propose the D3PG algorithm for more efficient and interpretable learning in robotic control. By decomposing the whole learning robot into multiple learning agents, the high-dimensional space can be reduced into low-dimensional space, thus reducing the overall complexity to solve the problem. We have shown that D3PG can achieve competitive or significantly improved performance compared to the existing DRL algorithms. Another advantage of D3PG is that it can explain the importance of each agent when it performs a certain task. In the future, we plan to investigate in the optimal K value automatically in D3PG-PCG.

A   Appendix

A.1   Appendix

A.2   MuJoCo Platform

Below we provide some specifications for the states, actions and rewards of the four robot environments in MuJoCo.

Swimmer. The swimmer is a planar robot with 3 links and 2 actuated joints. Fluid is simulated through viscosity forces, which apply drag on each link, allowing the swimmer to move forward. The 8-dim observation includes the joint angles and velocities, and the coordinates of the center of mass. The reward is given by $r(s, a) = v_x - 0.005 \cdot \|a\|_2^2$, where v_x is the forward velocity. No termination condition is applied.

Hopper. The hopper is a planar robot with 4 rigid links, corresponding to the torso, upper leg, lower leg, and foot, along with 3 actuated joints. The 11-dim observation includes joint angles, joint velocities, the coordinates of the center of mass, and the constraint forces. The reward is given by $r(s, a) = v_x - 0.005 \cdot \|a\|_2^2 + 1$, where the last term is a bonus for being "alive". The episode is terminated when $z_{body} < 0.7$, where z_{body} is the z-coordinate of the body, or when $|\theta_y| < 0.2$, where θ_y is the forward pitch of the body.

Walker. The walker is a planar biped robot consisting of 7 links, corresponding to two legs and a torso, along with 6 actuated joints. The 17-dim observation includes joint angles, joint velocities, and the coordinates of center of mass. The reward is given by $r(s, a) = v_x - 0.005 \cdot \|a\|_2^2$. The episode is terminated when $z_{body} < 0.8$, or $z_{body} > 2.0$, or $|\theta_y| > 1.0$.

Half-Cheetah. The half-cheetah is a planar biped robot with 9 rigid links, including two legs and a torso, along with 6 actuated joints. The 17-dim state includes joint angles, joint velocities, and the coordinates of the center of mass. The reward $r(s, a) = v_x - 0.005 \cdot \|a\|_2^2$. No termination condition is applied.

References

1. Bu, L., Babu, R., De Schutter, B., et al.: A comprehensive survey of multiagent reinforcement learning. IEEE TSMC Part C **38**(2), 156–172 (2008)
2. Busoniu, L., De Schutter, B., Babuska, R.: Decentralized reinforcement learning control of a robotic manipulator. In: 2006 9th ICARCV, pp. 1–6 (2006)
3. Dong, Y., Yu, C., Weng, P., Maustafa, A., Cheng, H., Ge, H.: Decomposed deep reinforcement learning for robotic control. In: Proceedings of the 19th International Conference on Autonomous Agents and MultiAgent Systems, pp. 1834–1836 (2020)
4. Duan, Y., Chen, X., Houthooft, R., Schulman, J., Abbeel: Benchmarking deep reinforcement learning for continuous control. In: ICML, pp. 1329–1338 (2016)
5. Dziomin, U., Kabysh, A., Golovko, V., Stetter, R.: A multi-agent reinforcement learning approach for the efficient control of mobile robot. In: 2013 IEEE 7th IDAACS, vol. 2, pp. 867–873 (2013)
6. Foerster, J., Assael, I.A., de Freitas, N., Whiteson, S.: Learning to communicate with deep multi-agent reinforcement learning. In: NIPS, pp. 2137–2145 (2016)
7. Foerster, J.N., Farquhar, G., Afouras, T., et al.: Counterfactual multi-agent policy gradients. In: AAAI, pp. 7254–7264 (2018)
8. Guestrin, C., Lagoudakis, M., Parr, R.: Coordinated reinforcement learning. In: ICML, vol. 2, pp. 227–234 (2002)
9. Gupta, J.K., Egorov, M., Kochenderfer, M.: Cooperative multi-agent control using deep reinforcement learning. In: Sukthankar, G., Rodriguez-Aguilar, J.A. (eds.) AAMAS 2017. LNCS (LNAI), vol. 10642, pp. 66–83. Springer, Cham (2017). https://doi.org/10.1007/978-3-319-71682-4_5
10. Iqbal, S., Sha, F.: Actor-attention-critic for multi-agent reinforcement learning. arXiv preprint arXiv:1810.02912 (2018)
11. Jiang, J., Lu, Z.: Learning attentional communication for multi-agent cooperation. In: NIPS, pp. 7254–7264 (2018)
12. Kabysh, A., Golovko, V., Lipnickas, A.: Influence learning for multi-agent system based on reinforcement learning. Int. J. Comput. **11**(1), 39–44 (2014)

13. Kok, J.R., Vlassis, N.: Collaborative multiagent reinforcement learning by payoff propagation. J. Mach. Learn. Res. **7**(Sep), 1789–1828 (2006)
14. Leottau, D.L., Ruiz-del-Solar, J., MacAlpine, P., Stone, P.: A study of layered learning strategies applied to individual behaviors in robot soccer. In: Almeida, L., Ji, J., Steinbauer, G., Luke, S. (eds.) RoboCup 2015. LNCS (LNAI), vol. 9513, pp. 290–302. Springer, Cham (2015). https://doi.org/10.1007/978-3-319-29339-4_24
15. Leottau, D.L., Ruiz-del Solar, J., Babuška, R.: Decentralized reinforcement learning of robot behaviors. Artif. Intell. **256**, 130–159 (2018)
16. Leottau, D.L., Ruiz-del Solar, J.: An accelerated approach to decentralized reinforcement learning of the ball-dribbling behavior. In: Workshops at the Twenty-Ninth AAAI (2015)
17. Li, Y.: Deep reinforcement learning. arXiv preprint arXiv:1810.06339 (2018)
18. Lillicrap, T.P., Hunt, J.J., Pritzel, A., et al.: Continuous control with deep reinforcement learning. arXiv preprint arXiv:1509.02971 (2015)
19. Lowe, R., Wu, Y., Tamar, A., et al.: Multi-agent actor-critic for mixed cooperative-competitive environments. In: NIPS, pp. 6379–6390 (2017)
20. Mao, H., Zhang, Z., et al.: Modelling the dynamic joint policy of teammates with attention multi-agent DDPG. In: AAMAS, pp. 1108–1116 (2019)
21. Martin, J.A., De Lope, H., et al.: A distributed reinforcement learning architecture for multi-link robots. In: 4th ICINCO, vol. 192, p. 197 (2007)
22. Matignon, L., Laurent, G.J., Le Fort-Piat, N.: Design of semi-decentralized control laws for distributed-air-jet micromanipulators by reinforcement learning. In: 2009 IROS, pp. 3277–3283 (2009)
23. Mnih, V., Badia, A.P., et al.: Asynchronous methods for deep reinforcement learning. In: ICML, pp. 1928–1937 (2016)
24. Peng, P., Yuan, Q., et al.: Multiagent bidirectionally-coordinated nets for learning to play StarCraft combat games, p. 2. arXiv preprint arXiv:1703.10069 (2017)
25. Peters, J., Schaal, S.: Natural actor-critic. Neurocomputing **71**(7–9), 1180–1190 (2008)
26. Rashid, T., Samvelyan, M., et al.: QMIX: monotonic value function factorisation for deep multi-agent reinforcement learning. In: ICML, pp. 4292–4301 (2018)
27. Sanchez-Gonzalez, A., Heess, N., et al.: Graph networks as learnable physics engines for inference and control. arXiv preprint arXiv:1806.01242 (2018)
28. Schulman, J., Levine, S., Abbeel, P., et al.: Trust region policy optimization. In: ICML, pp. 1889–1897 (2015)
29. Schulman, J., Moritz, P., et al.: High-dimensional continuous control using generalized advantage estimation. arXiv preprint arXiv:1506.02438 (2015)
30. Schulman, J., Wolski, F., et al.: Proximal policy optimization algorithms. arXiv preprint arXiv:1707.06347 (2017)
31. Sunehag, P., Lever, G., et al.: Value-decomposition networks for cooperative multi-agent learning based on team reward. In: AAMAS, pp. 2085–2087 (2018)
32. Sutton, R.S., Barto, A.G.: Reinforcement Learning: An Introduction. MIT Press, Cambridge (2018)
33. Todorov, E., Erez, T., Tassa, Y.: MuJoCo: a physics engine for model-based control. In: 2012 IROS, pp. 5026–5033 (2012)
34. Troost, S., Schuitema, E., Jonker, P.: Using cooperative multi-agent Q-learning to achieve action space decomposition within single robots. In: 1st ERLARS, p. 23 (2008)
35. Vaswani, A., Shazeer, N., et al.: Attention is all you need. In: NIPS, pp. 5998–6008 (2017)

36. Wang, T., Liao, R., et al.: NerveNet: learning structured policy with graph neural networks (2018)
37. Williams, R.J.: Simple statistical gradient-following algorithms for connectionist reinforcement learning. Mach. Learn. **8**(3–4), 229–256 (1992). https://doi.org/10.1007/BF00992696
38. Yu, C., Wang, D., Ren, J., Ge, H., Sun, L.: Decentralized multiagent reinforcement learning for efficient robotic control by coordination graphs. In: Geng, X., Kang, B.-H. (eds.) PRICAI 2018. LNCS (LNAI), vol. 11012, pp. 191–203. Springer, Cham (2018). https://doi.org/10.1007/978-3-319-97304-3_15

Lyapunov-Based Reinforcement Learning for Decentralized Multi-agent Control

Qingrui Zhang[1], Hao Dong[2], and Wei Pan[3(✉)]

[1] School of Aeronautics and Astronautics, Sun Yat-Sen University,
Guangzhou, China
qingrui.zhang@tudelft.nl
[2] Center on Frontiers of Computing Studies, Peking University, Beijing, China
hao.dong@pku.edu.cn
[3] Department of Cognitive Robotics, Delft University of Technology,
Delft, The Netherlands
wei.pan@tudelft.nl

Abstract. Decentralized multi-agent control has broad applications, ranging from multi-robot cooperation to distributed sensor networks. In decentralized multi-agent control, systems are complex with unknown or highly uncertain dynamics, where traditional model-based control methods can hardly be applied. Compared with model-based control in control theory, deep reinforcement learning (DRL) is promising to learn the controller/policy from data without the knowing system dynamics. However, to directly apply DRL to decentralized multi-agent control is challenging, as interactions among agents make the learning environment non-stationary. More importantly, the existing multi-agent reinforcement learning (MARL) algorithms cannot ensure the closed-loop stability of a multi-agent system from a control-theoretic perspective, so the learned control polices are highly possible to generate abnormal or dangerous behaviors in real applications. Hence, without stability guarantee, the application of the existing MARL algorithms to real multi-agent systems is of great concern, e.g., UAVs, robots, and power systems, etc. In this paper, we aim to propose a new MARL algorithm for decentralized multi-agent control with a stability guarantee. The new MARL algorithm, termed as a multi-agent soft-actor critic (MASAC), is proposed under the well-known framework of "centralized-training-with-decentralized-execution". The closed-loop stability is guaranteed by the introduction of a stability constraint during the policy improvement in our MASAC algorithm. The stability constraint is designed based on Lyapunov's method in control theory. To demonstrate the effectiveness, we present a multi-agent navigation example to show the efficiency of the proposed MASAC algorithm.

Keywords: Multi-agent reinforcement learning · Lyapunov stability · Decentralized control · Collective robotic systems

© Springer Nature Switzerland AG 2020
M. E. Taylor et al. (Eds.): DAI 2020, LNAI 12547, pp. 55–68, 2020.
https://doi.org/10.1007/978-3-030-64096-5_5

1   Introduction

Multi-agent system control has intrigued researchers from both industrial and academic communities for decades, due to its prospect in broad applications, such as formation flight of unmanned aerial vehicles (UAVs) [43,44], coordination of multi-robots [4,36], flocking/swarm control [33,40], distributed sensor networks [32], large-scale power systems [17], traffic and transportation systems [6], etc. Control of a multi-agent system can be achieved in either a centralized or a decentralized manner. However, a multi-agent system with many subsystems has high state and action dimensions that will dramatically increase the design complexity and computational burdens of a single centralized controller [3]. In many applications, every agent of a multi-agent system only has local control capability with access to local observations, e.g., the cooperation of multiple vehicles [13]. The lack of global control capability and information excludes the possibility of centralized control. Besides, centralized control tends to be less reliable. If the central controller fails, the entire system will break down. As an alternative, decentralized control is capable of handling all the above issues.

Decentralized multi-agent control has been extensively studied [9,34,35,41]. With the assumption that the agents' dynamics are known and linear, many model-based control algorithms have been proposed for different tasks [9,36,41]. In control theory, those model-based algorithms can ensure closed-loop stability if a multi-agent system satisfies all the assumptions. The state trajectories of a multi-agent system under a model-based control algorithm will always stay close to or even converge to an equilibrium point [24]. However, in most applications, agent dynamics are nonlinear, complicated, and highly uncertain, e.g., robotic systems, UAVs, and power systems. Assumptions made by model-based control algorithms can be barely satisfied in real life. Therefore, model-based control algorithms are restrictive, though theoretically sound.

Compared with model-based control, deep reinforcement learning (DRL) is more promising for the decentralized multi-agent control for complicated non-linear dynamical systems, as it can learn controller/policy from samples without using much model information [7,8,12,26,39,45,46]. Recently, deep RL has obtained significant success in applying to a variety of complex single-agent control problems [2,19,25,30]. However, it is more challenging to apply deep RL to decentralized multi-agent control. In multi-agent reinforcement learning (MARL), agents seek the best responses to other agents' policies. The policy update of an agent will affect the learning targets of other agents. Such interactions among agents make MARL training non-stationary, thus influencing the learning convergence. To resolve the non-stationary issue, a "centralized-training-with-decentralized-execution" mechanism was employed, based on which a number of MARL algorithms have been proposed, e.g., MAD-DPG [29], COMA [14], mean-field MARL [42], MATD3 [1], and MAAC [22], etc. Unfortunately, the existing MARL algorithms can not ensure the closed-loop stability for a multi-agent system, while stability is the foremost concern for the control of any dynamical systems. It is highly possible that learned control polices will generate abnormal or risky behaviors in real applications. From a control perspective, the learned control policies fail to stabilize a multi-agent system, so they cannot be applied to safety-critical scenarios, e.g., formation flight of UAVs.

In this paper, we propose MARL algorithms for decentralized multi-agent control with a stability guarantee. A multi-agent soft actor-critic (MASAC) algorithm is developed based on the well-known "centralized-training-with-decentralized-execution" scheme. The interactions among agents are characterized using graph theory [11]. Besides, a stability-related constraint is introduced to the policy improvement to ensure the closed-loop stability of the learned control policies. The stability-related constraint is designed based on the well-known Lyapunov's method in control theory which is a powerful tool for the design of a controller to stabilize the complex nonlinear systems with stability guarantee [24].

Contributions: The contributions of this paper can be summarized as follows.

1. For the first time, a Lyapunov-based multi-agent soft actor-critic algorithm is developed for decentralized control problems based on the "centralized-training-with-decentralized-execution" to guarantee the stability of a multi-agent system.
2. Theoretical analysis is presented on the design of a stability constraint using Lyapunov's method.

2 Preliminaries

2.1 Networked Markov Game

Interactions among N agents are characterized using an undirected graph $\mathcal{G} = \langle \mathcal{I}, \mathcal{E} \rangle$, where $\mathcal{I} := \{1, \ldots, N\}$ represents the set of N agents and $\mathcal{E} \subseteq \mathcal{I} \times \mathcal{I}$ denotes the interactions among agents. If an agent i is able to interact with an agent j with $j \neq i$ and $i, j \in \mathcal{I}$, there exists an edge $(i, j) \in \mathcal{E}$, and agent j is called a neighbor of agent i. For an undirected graph, $(j, i) \in \mathcal{E}$ if $(i, j) \in \mathcal{E}$. The neighborhood of agent i is denoted by $\mathcal{N}_i := \{\forall j \in \mathcal{I} | (i, j) \in \mathcal{E}\}$. Assume the undirected graph is fully connected, so there exists a path from each node $i \in \mathcal{I}$ to any other nodes $j \in \mathcal{I}$ [11,16]. If an undirected graph is strongly connected, information could eventually be shared among all agents via the communication graph.

A networked Markov game with N agents is denoted by a tuple, $\mathcal{MG} := \langle \mathcal{G}, \mathcal{S}, \mathcal{A}, \mathcal{P}, r, \gamma \rangle$, where $\mathcal{G} := \langle \mathcal{I}, \mathcal{E} \rangle$ is the communication graph among N agents, $\mathcal{S} := \bigcup_{i=1}^{N} \mathcal{S}_i$ is the entire environment space with $\mathcal{S}_i$ the local state space for agent $i \in \mathcal{I}$, $\mathcal{A} := \bigcup_{i=1}^{N} \mathcal{A}_i$ denotes the joint action space with $\mathcal{A}_i$ the local action space for agent $i \in \mathcal{I}$, $\mathcal{P} := \mathcal{S} \times \mathcal{A} \times \mathcal{S} \rightarrow \mathbb{R}$ specifies the state transition probability function, and $r := \mathcal{S} \times \mathcal{A} \rightarrow \mathbb{R}$ represents the global reward function of the entire multi-agent system. The global transition probability can, therefore, be denoted by $\mathcal{P}(s_{t+1}|s_t, a_t)$. The joint action of N agents is $a = \{a_1, \ldots, a_N\}$ where a_i denotes the action of an agent $i \in \mathcal{I}$. Accordingly, the joint policy is defined to be $\boldsymbol{\pi} = \{\pi_1, \ldots, \pi_N\}$ where π_i ($\forall i \in \mathcal{I}$) are local policies for an agent i. Hence, the global policy for the entire multi-agent system is defined to be $\pi(\boldsymbol{a}|\boldsymbol{s}) = \prod_{i=1}^{N} \pi_i(a_i|s_i)$. Assume each agent i can only obtain a local observation $s_i \in \mathcal{S}_i$ (e.g. its own states and state information of its neighbors) to make decisions at the execution.

For any given initial global state $\boldsymbol{s}_0$, the global expected discounted return following a joint policy $\boldsymbol{\pi}$ is given by

$$V(\boldsymbol{s}_t) = \sum_{t=0}^{\infty} \mathbb{E}_{(\boldsymbol{s}_t, \boldsymbol{a}_t) \sim \rho_\pi} \left[\gamma^t r(\boldsymbol{s}_t, \boldsymbol{a}_t) | \boldsymbol{s}_0 \right] \tag{1}$$

where γ is a discount factor, V is the global value function and ρ_π is the state-action marginals of the trajectory distribution induced by a global policy $\boldsymbol{\pi}$. The global action-value function (a.k.a. Q-function) of the entire system is

$$Q(\boldsymbol{s}_t, \boldsymbol{a}_t) = r(\boldsymbol{s}_t, \boldsymbol{a}_t) + \gamma \mathbb{E}_{\boldsymbol{s}_{t+1}} [V(\boldsymbol{s}_{t+1})] \tag{2}$$

2.2   Soft Actor-Critic Algorithm

In this paper, the soft actor-critic (SAC) algorithm will be used for the design of the multi-agent reinforcement learning algorithm. The SAC algorithm belongs to off-policy RL that is more sample efficient than on-policy RL methods [39], such as the trust region policy optimization (TRPO) [37] and the proximal policy optimization (PPO) [38]. In SAC, an expected entropy of the policy $\boldsymbol{\pi}$ is added to the value functions (1) and (2) to regulate the exploration performance at the training stage [19,47]. The inclusion of the entropy term makes the SAC algorithm exceed both the efficiency and final performance of deep deterministic policy gradient (DDPG) [18,27]. With the inclusion of the expected entropy, the action value function (1) to be maximized for a multi-agent system will turn into

$$V(\boldsymbol{s}_t) = \sum_{t=0}^{\infty} \mathbb{E}_{(\boldsymbol{s}_t, \boldsymbol{a}_t) \sim \rho_\pi} \left[\gamma^t \left(r(\boldsymbol{s}_t, \boldsymbol{a}_t) + \alpha \mathcal{H}(\boldsymbol{\pi}(\cdot|\boldsymbol{s}_t)) \right) | \boldsymbol{s}_0 \right] \tag{3}$$

where α is the temperature parameter used to control the stochasticity of the policy by regulating the relative importance of the entropy term against the reward, and $\mathcal{H}(\boldsymbol{\pi}(\cdot|\boldsymbol{s}_t)) = -\mathbb{E}_\pi \left[\log(\boldsymbol{\pi}(\cdot|\boldsymbol{s}_t)) \right]$ is the entropy of the policy $\boldsymbol{\pi}$. Accordingly, a modified Bellman backup operator is defined as

$$\mathcal{T}^\pi Q(\boldsymbol{s}_t, \boldsymbol{a}_t) = r(\boldsymbol{s}_t, \boldsymbol{a}_t) + \gamma \mathbb{E}_{\boldsymbol{s}_{t+1}} [V(\boldsymbol{s}_{t+1})] \tag{4}$$

where $V(\boldsymbol{s}_t) = \mathbb{E}_{\boldsymbol{a}_t \sim \pi} [Q(\boldsymbol{s}_t, \boldsymbol{a}_t) - \alpha \log(\boldsymbol{\pi}(\boldsymbol{a}_t|\boldsymbol{s}_t))]$.

2.3   Lyapunov Stability in Control Theory

A dynamical system is called stable, if its state trajectory starting in vicinity to an equilibrium point will stay near the equilibrium point all the time. Stability is a crucial concept for the control and safety of any dynamical systems. Lyapunov stability theory provides a powerful means of stabilizing unstable dynamical systems using feedback control. The idea is to select a suitable Lyapunov function and force it to decrease along the trajectories of the system. The resulting system will eventually converge to its equilibrium. Lyapunov stability of dynamic systems at a fixed policy $\pi(a|s)$ is given by Lemma 1.

Lemma 1 (Lyapunov stability). [24] *Suppose a system is denoted by a nonlinear mapping $s_{t+1} = f(s_t, \pi(a_t|s_t))$. Let $L(s_t)$ be a continuous function such that $L(s_T) = 0$, $L(s_t) > 0$ ($\forall s_t \in \Omega$ & $s_t \neq s_T$), and $L(s_{t+1}) - L(s_t) \leq 0$ ($\forall s_t \in \Omega$) where s_T is an equilibrium state and Ω is a compact state space. Then the system $s_{t+1} = f(s_t, \pi(a_t|s_t))$ is stable around s_T and $L(s_t)$ is a Lyapunov function. Furthermore, if $L(s_{t+1}) - L(s_t) < 0$ ($\forall s_t \in \Omega$), the system is asymptotically stable around s_T.*

Note that maximizing the objective function (1) or (2) doesn't necessarily result in a policy stabilizing a dynamical system.

3 Multi-agent Reinforcement Learning with Lyapunov Stability Constraint

In this section, we will first develop a MARL algorithm based on the SAC algorithm by following a similar idea as the multi-agent deterministic policy gradient descent (MADDPG) [29]. The proposed algorithm is termed as Multi-Agent Soft Actor-Critic algorithm (MASAC). The proposed MASAC algorithm is thereafter enhanced by incorporating a carefully designed Lyapunov constraint.

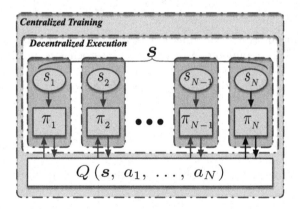

Fig. 1. Centralized training with decentralized execution

3.1 Multi-agent Soft Actor-Critic Algorithm

The crucial concept behind the MASAC is the so-called *"centralized training with decentralized execution"* shown in Fig. 1. A centralized critic using global information is employed at the training stage, while each agent uses their own independent policy taking local observations as inputs. Hence, at the training stage, agents share their rewards with all the other agents for the calculation of the central critic. In MASAC, it is expected to maximize the entropy-regularized objective function introduced in (3).

In decentralized control, agents make decisions based on their local observations [3,13,23,34,35]. Hence, their polices are assumed to be independent of one another, i.e., $\pi(a_t|s_t) = \prod_{i=1}^{N} \pi_i(a_i|s_i)$. Hence, the entropy of the joint policy $\pi(a_t|s_t)$ in (3) is

$$\mathcal{H}(\pi) = -\sum_i^N \mathbb{E}_{\pi_i}\left[\log(\pi_i)\right] = \sum_i^N \mathcal{H}(\pi_i) \tag{5}$$

where $\mathcal{H}(\pi_i)$ represent the entropy of each local policy π_i.

The entire algorithm is divided into policy evaluation and policy improvement. In the policy evaluation step, we will repeatedly apply the modified Bellman backup operator (4) to the Q-value of a fixed joint policy. Let the centralized Q-value for the multi-agent system be parameterized by θ. The critic neural network parameter θ is trained to minimize the following Bellman residual.

$$J_Q(\theta) = \mathbb{E}_{(s_t,a_t)\sim\mathcal{D}}\left[\frac{1}{2}\left(Q_\theta(s_t,a_t) - r(s_t,a_t)\right.\right.$$
$$\left.\left. - \gamma\mathbb{E}_{s_{t+1}}\left[V_{\bar\theta}(s_{t+1}) + \alpha\sum_i^N \mathcal{H}(\pi_{\phi_i})\right]\right)^2\right] \tag{6}$$

In the optimization, the value function is replaced by the Q-value function. Therefore, the critic parameters are optimized by stochastic gradient descent as

$$\nabla_\theta J_Q(\theta) = \mathbb{E}_{(s_t,a_t)\sim\mathcal{D}}\left[\nabla_\theta Q_\theta(s_t,a_t)\delta_Q\right] \tag{7}$$

where

$$\delta_Q = Q_\theta(s_t,a_t) - r - \gamma Q_{\bar\theta}(s_{t+1},a_{t+1}) + \gamma\alpha\sum_i^N \log\pi_{\phi_i} \tag{8}$$

In policy improvement, the policy is updated according to

$$\pi* = arg\min_{\pi'\in\Pi}\mathbb{E}_{\pi_i}\left[\alpha\sum_i^N \log(\pi_i) - Q(s_t,a_t)\right] \tag{9}$$

where $\pi^* = \{\pi_1^*, \ldots, \pi_N^*\}$ is the optimal joint policy. Assume the policy of agent i is parameterized by ϕ_i, $\forall i = 1,\ldots,N$. According to (9), the policy parameters ϕ_i, $\forall i = 1,\ldots,N$ are trained to minimize

---

Algorithm 1. Multi-agent soft actor-critic algorithm

Initialize parameters θ^1, θ^2 and ϕ_i $\forall i \in \mathcal{I}$
$\bar{\theta}^1 \leftarrow \theta^1$, $\bar{\theta}^2 \leftarrow \theta^2$, $\mathcal{D} \leftarrow \emptyset$
repeat
    for each environment step **do**
        $a_{i,t} \sim \pi_{\phi_i}(a_{i,t}|s_{i,t})$, $\forall i \in \mathcal{I}$
        $s_{t+1} \sim \mathcal{P}_i(s_{t+1}|s_t, \boldsymbol{a}_t)$, where $\boldsymbol{a}_t = \{a_{1,t}, \ldots, a_{N,t}\}$
        $\mathcal{D} \leftarrow \mathcal{D} \bigcup \{s_t, \boldsymbol{a}_t, r(s_t, \boldsymbol{a}_t,), s_{t+1}\}$
    end for
    for each gradient update step **do**
        Sample a batch of data, $\mathcal{B}$, from $\mathcal{D}$
        $\theta^j \leftarrow \theta^j - \iota_Q \nabla_\theta J_Q(\theta^j)$, $j = 1, 2$
        $\phi_i \leftarrow \phi_i - \iota_\pi \nabla_{\phi_i} J_{\pi_i}(\phi_i)$, $\forall i \in \mathcal{I}$
        $\alpha \leftarrow \alpha - \iota_\alpha \nabla_\alpha J_\alpha(\alpha)$
        $\bar{\theta}^j \leftarrow \tau\theta^j + (1 - \tau)\bar{\theta}^j$, $j = 1, 2$
    end for
until convergence

$$J_\pi(\phi) \simeq \mathbb{E}_{(s_t, a_t) \sim \mathcal{D}}\left(\mathbb{E}_{\pi_\phi}\left(\alpha \sum_i^N \log(\pi_{\phi_i}) - Q_\theta(s_t, a_t))\right)\right) \tag{10}$$

where $\phi = \{\phi_1, \ldots, \phi_N\}$ and $\pi_\phi = \{\pi_{\phi_1}, \ldots, \pi_{\phi_N}\}$. In terms of the stochastic gradient descent, each agent's policy parameter ϕ_i will be updated according to

$$\nabla_{\phi_i} J_\pi(\phi) \simeq \mathbb{E}_{(s_t, a_t) \sim \mathcal{D}}\left[\left(\nabla_{a_i} \log \pi_{\phi_i} - \nabla_{a_i} Q_{\theta_i}(s_t, a_t, \bar{a}_t)\right)\nabla_{\phi_i} a_{\phi_i}\right.$$
$$\left. + \nabla_{\phi_i} \log \pi_{\phi_i}\right] \tag{11}$$

The temperature parameter α will be updated based on (12).

$$J_{\alpha_i} = \mathbb{E}_\pi\left[-\alpha \sum_i^N \log \pi_i - \alpha\bar{\mathcal{H}}\right] \tag{12}$$

The MASAC algorithm is summarized in Algorithm 1. The final MASAC algorithm uses two soft Q-functions to mitigate the estimation bias in the policy improvement and further increase the algorithm performance [1,15,21].

3.2 Lyapunov Stability Constraint

Qualitatively, stability implies that the states of a system will be at least bounded and stay close to an equilibrium state for all the time. The existing MARL algorithms, including the proposed MASAC algorithm in Sect. 3.1, can find an optimal policy that can maximize either state or action-value functions. However, they do not necessarily produce a policy that ensures the stability of a system. In this section, we offer a possible solution to incorporate Lyapunov stability as a constraint in the optimization of MASAC.

A Lyapunov function candidate can be constructed based on cost functions $c(s, \pi) \geq 0$ with $c(s_T, \pi(a_T|s_T)) = 0$ and s_T the target/equilibrium state [5,10]. One possible choice of the Lyapunov function candidate is an accumulated cost in a finite time horizon, e.g., model predictive control (MPC) [31]. Before the introduction of the Lyapunov stability constraint, we make several assumptions on both the cost functions to be designed and the system dynamics of interest. The assumption on the cost function is given as follows.

Assumption 1. *The cost function $c(s, \pi(\cdot|s))$ is bounded $\forall s \in \Omega$ and Lipschitz continuous with respect to s, namely $\|c(s_1, \pi(\cdot|s_1)) - c(s_2, \pi(\cdot|s_2))\|_2 \leq l_c \|s_1 - s_2\|_2$ where $l_c > 0$ is a Lipschitz constant.*

Assumption 1 could ensure the Lyapunov function candidate to be bounded and Lipschitz continuous for the state s, if we choose it to be an accumulated cost in a finite time horizon. Hence, we could further assume the Lyapunov function candidate is Lipschitz continuous with the state s on a compact set Ω with l_L the Lipschitz constant.

Since we are interested in the decentralized control problem of multiple agents with deterministic dynamics, the following assumption on the physical dynamics is made.

Assumption 2. *Consider a deterministic, discrete-time agent system $s_{t+1} = f(s_t, a_t)$. The nonlinear dynamics f is Lipschitz continuous with respect to a_t, namely $\|f(s_t, a_t^2) - f(s_t, a_t^1)\|_2 \leq l_f \|a_t^2 - a_t^1\|_2$ where $l_f > 0$ is a Lipschitz constant.*

According to the existence and uniqueness theorem, the local Lipschitz condition is a common assumption for deterministic continuous systems.

According to Lemma 1, the state s_{t+1} is needed to evaluate the stability of a system under a fixed policy, but s_{t+1} is not available in general. In Theorem 1, we show that it is possible to evaluate the stability of a new policy π_{new}, if we already have a feasible policy π_{old} associated with a Lyapunov function $L(s)$. Here, a feasible policy implies that a system is stable and that a Lyapunov function exists.

Theorem 1. *Consider a system $s_{t+1} = f(s_t, a_t)$ Suppose Assumptions 1 and 2 hold. Let π_{old} be a feasible policy for data collection and $L_{\pi_{old}}(s)$ is the Lyapunov function. A new policy π_{new} will also be a feasible policy under which the system is stable, if there exists*

$$L_{\pi_{old}}(s_{t+1}) + l_L l_f \|a_t^{\pi_{new}} - a_t^{\pi_{old}}\|_2 - L_{\pi_{old}}(s_t) \leq 0 \tag{13}$$

where $\forall s_t \in \Omega$, $(s_t, a_t^{\pi_{old}}, s_{t+1})$ is a tuple from the policy π_{old}, l_L and l_f are Lipschitz constants of the Lyapunov function and system dynamics, respectively.

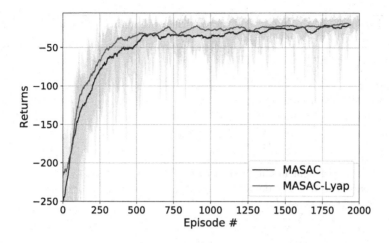

Fig. 2. Learning curves of the rendezvous experiment (40 steps per episode)

Theorem 1 requires all the states need to be visited to evaluate the stability of a new policy. Unfortunately, it is impossible to visit an infinite number of states. However, Theorem 1 still shows a potential way to use historical samples for the old policies to evaluate the current policy. Based on Theorem 1, we are able to add a Lyapunov constraint similar to (13) in the policy gradient of each agent for DC of multi-agent systems. With the inclusion of the Lyapunov constraint (13) [5, 20], the objective function (10) is rewritten as

$$J_\pi(\phi) \simeq \mathbb{E}_{(s_t,a_t)\sim\mathcal{D}}\left(\mathbb{E}_{\pi_\phi}\left(\alpha\sum_i^N \log(\pi_{\phi_i}) - Q_\theta(s_t,a_t)) + \beta\Delta L_\phi\right)\right) \qquad (14)$$

where $\beta \in [0,1]$ and $\Delta L_\psi = L_i(s_{t+1},a_{t+1}) + l_L l_f\|a_\phi - a_t\|_2 - L_i(s_t,a_t) + \beta c_\pi(s_t)$.

At training, the Lyapunov functions $L(s_t)$ will be parameterized by ψ which is trained to minimize

$$J_L(\psi) = \mathbb{E}_{(s_t,a_t)\sim\mathcal{D}}\left[\frac{1}{2}(L_\psi(s_t,a_t) - L_{target})^2\right]$$

where $L_{target} = \sum_{t=0}^T c(s_t,a_t)$ with T denoting a finite time horizon as in model predictive control. The modified robust multi-agent reinforcement learning algorithm is summarized in Algorithm 2.

4   Experiment

In this section, we will evaluate our proposed algorithms in a well-known application of multi-agent systems called "rendezvous" [28]. In the "rendezvous" problem, all agents starting from different locations are required to meet at the same

---

Algorithm 2. Multi-agent soft actor-critic algorithm with a Lyapunov constraint

---

Initialize parameters θ^1, θ^2 and ϕ_i $\forall i \in \mathcal{I}$
$\bar{\theta}^1 \leftarrow \theta^1$, $\bar{\theta}^2 \leftarrow \theta^2$, $\mathcal{D} \leftarrow \emptyset$
repeat
  for each environment step **do**
    $a_{i,t} \sim \pi_{\phi_i}(a_{i,t}|s_{i,t})$, $\forall i \in \mathcal{I}$
    $s_{t+1} \sim \mathcal{P}_i(s_{t+1}|s_t, a_t)$, where $a_t = \{a_{1,t}, \ldots, a_{N,t}\}$
    $\mathcal{D} \leftarrow \mathcal{D} \bigcup \{s_t, a_t, r(s_t, a_t), s_{t+1}\}$
  end for
  for each gradient update step **do**
    Sample a batch of data, $\mathcal{B}$, from $\mathcal{D}$
    $\theta^j \leftarrow \theta^j - \iota_Q \nabla_{\theta_i} J_{Q_i}(\theta^j)$, $j = 1, 2$
    $\phi_i \leftarrow \phi_i - \iota_\pi \nabla_{\phi_i} J_\pi(\phi)$, $\forall i \in \mathcal{I}$ from (14)
    $\alpha \leftarrow \alpha - \iota_\alpha \nabla_\alpha J_\alpha(\alpha)$
    $\bar{\theta}^j \leftarrow \tau \theta^j + (1 - \tau) \bar{\theta}^j$, $j = 1, 2$
    $\psi \leftarrow \psi - \iota_L \nabla_\psi J_L(\psi)$
  end for
until convergence

---

target location in the end. Only a subgroup of agents, which are called leaders, have access to the target location, while others need to learn to cooperate with others. In the experiments, both the critic and actor are represented suing fully connected multiple-layer perceptrons with two hidden layers. Each hidden Layer has 64 neurons with the 'ReLU' activation function. The learning rate for the actor network is chosen to be 0.0003, while the learning rate for the critic network is 0.003. To stabilize the training, learning rates decrease with a certain decay rate ($0.075^{0.0005}$ in the experiments). The Lyapunov neural network is approximated by an MLP with three hidden layers (64 neurons for the first two hidden layers, and 16 neurons for the last hidden layers). The batch size is selected to be 256. The parameter τ for soft updates of both actor and critic networks is picked to be 0.005. The discount factor γ is chosen to be 0.95.

The environment is built using the multi-agent environment used in [29]. The agent model in the environment in [29] is replaced by a high-order non-holonomic unicycle model which is widely used in robotics navigation. The agent dynamics are given as follows.

$$\begin{cases} \dot{x} = v \cos \psi \\ \dot{y} = v \sin \psi \\ \dot{\psi} = \omega \\ \dot{v} = a \\ \dot{\omega} = r \end{cases} \tag{15}$$

where x and y are the positions of the agent, ψ is the heading angle, v is the speed, and ω is the angular rate. The control actions for each agent are a and r, respectively.

Learning curves of both the MASAC and MASAC-Lyapunov are shown in Fig. 2. Although both the MASAC and MASAC-Lyapunov will converge, they have a different performance at evaluations. To further verify the performance, we evaluate both the MASAC and MASAC-Lyapunov for 500 times with agents' initial positions randomly generated. We define a criterion called "success rate" to compare the overall performance of the two algorithms. We think an evaluation episode is successful if all agents end up in the target location. The "success rate" is calculated by $\frac{\text{total number of successful episodes}}{total number of episodes at evaluation} \times 100$. The "success rate" of the two algorithms is shown in Fig. 3. With the inclusion of the Lyapunov constraint, we can increase the success rate of the tasks dramatically according to Fig. 3. Hence, the Lyapunov constraint will increase the stability performance of the learned policy, thereby resulting in a policy that is more likely to stabilize a system.

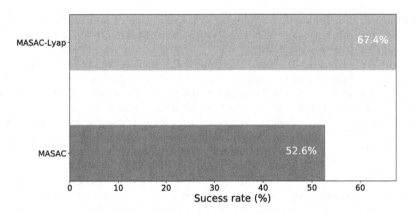

Fig. 3. Evaluation results of running the rendezvous experiment for 500 times using trained polices (success rate $= \frac{\text{total number of successful episodes}}{\text{total number of episodes at evaluation}} \times 100$)

5   Conclusion

In this paper, we studied MARL for data-driven decentralized control for multi-agent systems. We proposed a MASAC algorithm based on the "centralized-training-with-decentralized-execution'. We thereafter presented a feasible solution to combine Lyapunov's methods in control theory with MASAC to guarantee stability. The MASAC algorithm was modified accordingly by the introduction of a Lyapunov stability constraint. The experiment conducted in this paper demonstrated that the introduced Lyapunov stability constraint is important to design a policy to achieve better performance than our vanilla MASAC algorithm.

References

1. Ackermann, J., Gabler, V., Osa, T., Sugiyama, M.: Reducing overestimation bias in multi-agent domains using double centralized critics. arXiv preprint arXiv:1910.01465 (2019)
2. Andrychowicz, M., et al.: Learning dexterous in-hand manipulation. Int. J. Robot. Res. **30**(1), 3–20 (2020). https://doi.org/10.1177/0278364919887447
3. Bakule, L.: Decentralized control: an overview. Annu. Rev. Control **32**, 87–98 (2008). https://doi.org/10.1016/j.arcontrol.2008.03.004
4. van den Berg, J., Lin, M.C., Manocha, D.: Reciprocal velocity obstacles for real-time multi-agent navigation. In: Proceedings of the IEEE International Conference on Robotics and Automation (ICRA), Pasadena, CA, USA. IEEE, May 2008. https://doi.org/10.1109/ROBOT.2008.4543489
5. Berkenkamp, F., Turchetta, M., Schoellig, A.P., Krause, A.: Safe model-based reinforcement learning with stability guarantees. arXiv preprint arXiv:1705.08551 (2017)
6. Burmeister, B., Haddadi, A., Matylis, G.: Application of multi-agent systems in traffic and transportation. IEE Proc. Softw. Eng. **144**(1), 51–60 (1997). https://doi.org/10.1049/ip-sen:19971023
7. Chen, Y.F., Everett, M., Liu, M., How, J.P.: Socially aware motion planning with deep reinforcement learning. In: Proceedings of 2017 IEEE/RSJ International Conference on Intelligent Robots and Systems (IROS), Vancouver, BC, Canada, September 2017. https://doi.org/10.1109/IROS.2017.8202312
8. Chen, Y.F., Liu, M., Everett, M., How, J.P.: Decentralized non-communicating multiagent collision avoidance with deep reinforcement learning. In: Proceedings of 2017 IEEE International Conference on Robotics and Automation, Singapore, Singapore, June 2017. https://doi.org/10.1109/ICRA.2017.7989037
9. Cheng, Z., Zhang, H.T., Fan, M.C., Chen, G.: Distributed consensus of multi-agent systems with input constraints: a model predictive control approach. IEEE Trans. Circuits Syst. I Regul. Paper **62**(3), 825–834 (2015). https://doi.org/10.1109/TCSI.2014.2367575
10. Chow, Y., Nachum, O., Duenez-Guzman, E., Ghavamzadeh, M.: A Lyapunov-based approach to safe reinforcement learning. arXiv preprint arXiv:1805.07708 (2018)
11. Diestel, R.: Graph Theory, 2nd edn. Springer, New York (2000). https://doi.org/10.1007/978-3-662-53622-3
12. Everett, M., Chen, Y.F., How, J.P.: Collision avoidance in pedestrian-rich environments with deep reinforcement learning. arXiv preprint arXiv:1910.11689 (2019)
13. Feddema, J.T., Lewis, C., Schoenwald, D.A.: Decentralized control of cooperative robotic vehicles: theory and application. IEEE Trans. Robot. Autom. **18**(5), 852–864 (2002). https://doi.org/10.1109/TRA.2002.803466
14. Foerster, J.N., Farquhar, G., Afouras, T., Nardelli, N., Whiteson, S.: Counterfactual multi-agent policy gradients. arXiv preprint arXiv:1705.08926 (2017)
15. Fujimoto, S., van Hoof, H., Meger, D.: Addressing function approximation error in actor-critic methods. arXiv preprint arXiv:1802.09477 (2018)
16. Godsil, C., Royle, G.: Algebraic Graph Theory. Springer, New York (2000). https://doi.org/10.1007/978-1-4613-0163-9
17. Guo, Y., Hill, D.J., Wang, Y.: Nonlinear decentralized control of large-scale power systems. Automatica **36**(9), 1275–1289 (2000). https://doi.org/10.1016/S0005-1098(00)00038-8

18. Haarnoja, T., et al.: Soft actor-critic algorithms and applications. arXiv preprint arXiv:1812.05905 (2018)
19. Haarnoja, T., Zhou, A., Abbeel, P., Levine, S.: Soft actor-critic: off-policy maximum entropy deep reinforcement learning with a stochastic actor. arXiv preprint arXiv:1801.01290 (2018)
20. Han, M., Tian, Y., Zhang, L., Wang, J., Pan, W.: h_∞ model-free reinforcement learning with robust stability guarantee. arXiv preprint arXiv:1911.02875 (2019)
21. van Hasselt, H., Guez, A., Silver, D.: Deep reinforcement learning with double q-learning. arXiv preprint arXiv:1509.06461 (2015)
22. Iqbal, S., Sha, F.: Actor-attention-critic for multi-agent reinforcement learning. arXiv preprint arXiv:1810.0291 (2019)
23. Keviczky, T., Borrelli, F., Balas, G.J.: Decentralized receding horizon control for large scale dynamically decoupled systems. Automatica **42**, 2105–2115 (2006). https://doi.org/10.1016/j.automatica.2006.07.008
24. Khalil, H.K.: Nonlinear Systems, 3rd edn. Prentice Hall, Upper Saddle River (2001)
25. Kiran, B.R., et al.: Deep reinforcement learning for autonomous driving: a survey. arXiv preprint arXiv:2002.00444 (2020)
26. Levine, S.: Reinforcement learning and control as probabilistic inference: tutorial and review. arXiv preprint arXiv:1805.00909 (2018)
27. Lillicrap, T.P., et al.: Continuous control with deep reinforcement learning. arXiv preprint arXiv:1509.02971 (2015)
28. Lin, J., Morse, A., Anderson, B.: Lenient learners in cooperative multiagent systems. In: Proceedings of 42nd IEEE International Conference on Decision and Control (IEEE Cat. No. 03CH37475), Maui, HI, USA, December 2003. https://doi.org/10.1109/CDC.2003.1272825
29. Lowe, R., Wu, Y., Tamar, A., Harb, J., Abbeel, P., Mordatch, I.: Multiagent actor-critic for mixed cooperative-competitive environments. arXiv preprint arXiv:1706.02275 (2018)
30. Kalashnikov, D.: QT-Opt: scalable deep reinforcement learning for vision-based robotic manipulation. arXiv preprint arXiv:1806.10293 (2018)
31. Mayne, D., Rawlings, J., Rao, C., Scokaert, P.O.M.: Constrained model predictive control: stability and optimality. Automatica **36**(6), 789–814 (2000). https://doi.org/10.1016/S0005-1098(99)00214-9
32. Olfati-Saber, R., Shamma, J.: Consensus filters for sensor networks and distributed sensor fusion. In: Proceedings of 44-th IEEE International Conference on Decision and Control, Seville, Spain, December 2005. https://doi.org/10.1109/CDC.2005.1583238
33. Olfati-Saber, R., Fax, J.A., Murray, R.M.: Consensus and cooperation in networked multi-agent systems. Proc. IEEE **95**(1), (2007). https://doi.org/10.1109/JPROC.2006.887293
34. Ren, W.: Distributed leaderless consensus algorithms for networked Euler-Lagrange systems. Int. J. Control **82**(11), 2137–2149 (2009). https://doi.org/10.1080/00207170902948027
35. Ren, W., Beard, R.W.: Decentralized scheme for spacecraft formation flying via the virtual structure approach. J. Guid. Control Dyn. **27**(1), 706–716 (2004). https://doi.org/10.2514/1.9287
36. Rezaee, H., Abdollahi, F.: A decentralized cooperative control scheme with obstacle avoidance for a team of mobile robots. IEEE Trans. Industr. Electron. **61**(1), 347–354 (2014). https://doi.org/10.1109/TIE.2013.2245612

37. Schulman, J., Levine, S., Abbeel, P., Jordan, M., Moritz, P.: Trust region policy optimization. In: Proceedings of the 31st International Conference on Machine Learning, Lille, France, pp. 1889–1897, June 2015
38. Schulman, J., Wolski, F., Dhariwal, P., Radford, A., Klimov, O.: Proximal policy optimization algorithms. arXiv preprint arXiv:1707.06347 (2017)
39. Sutton, R.S., Barto, A.G.: Reinforcement Learning: An Introductions, 2nd edn. The MIT Press, Cambridge (2018)
40. Vásárhelyi, G., Virágh, C., Somorjai, G., Nepusz, T., Eiben, A.E., Vicsek, T.: Optimized flocking of autonomous drones in confined environments. Sci. Robot. **3**(20), (2018). https://doi.org/10.1126/scirobotics.aat3536
41. Wang, J., Xin, M.: Integrated optimal formation control of multiple unmanned aerial vehicles. IEEE Trans. Control Syst. Technol. **21**(5), 1731–1744 (2013). https://doi.org/10.1109/TCST.2012.2218815
42. Yang, Y., Luo, R., Li, M., Zhou, M., Zhang, W., Wang, J.: Mean field multi-agent reinforcement learning. arXiv preprint arXiv:1802.05438 (2018)
43. Zhang, Q., Liu, H.H.T.: Aerodynamic model-based robust adaptive control for close formation flight. Aerosp. Sci. Technol. **79**, 5–16 (2018). https://doi.org/10.1016/j.ast.2018.05.029
44. Zhang, Q., Liu, H.H.T.: UDE-based robust command filtered backstepping control for close formation flight. IEEE Trans. Industr. Electron. **65**(11), 8818–8827 (2018). https://doi.org/10.1109/TIE.2018.2811367. Accessed 12 Mar 2018
45. Zhang, Q., Pan, W., Reppa, V.: Model-reference reinforcement learning control of autonomous surface vehicles with uncertainties. arXiv preprint arXiv:2003.13839 (2020)
46. Zhang, Q., Pan, W., Reppa, V.: Model-reference reinforcement learning for collision-free tracking control of autonomous surface vehicles. arXiv preprint arXiv:2008.07240 (2020)
47. Ziebart, B.D.: Modeling Purposeful Adaptive Behavior with the Principle of Maximum Causal Entropy, December 2010. https://doi.org/10.1184/R1/6720692.v1. https://kilthub.cmu.edu/articles/Modeling_Purposeful_Adaptive_Behavior_with_the_Principle_of_Maximum_Causal_Entropy/6720692

Hybrid Independent Learning in Cooperative Markov Games

Roi Yehoshua[(✉)] and Christopher Amato

Northeastern University, Boston, MA 02115, USA
r.yehoshua@northeastern.edu

Abstract. Independent agents learning by reinforcement must overcome several difficulties, including non-stationarity, miscoordination, and relative overgeneralization. An independent learner may receive different rewards for the same state and action at different time steps, depending on the actions of the other agents in that state. Existing multi-agent learning methods try to overcome these issues by using various techniques, such as hysteresis or leniency. However, they all use the latest reward signal to update the Q function. Instead, we propose to keep track of the rewards received for each state-action pair, and use a hybrid approach for updating the Q values: the agents initially adopt an optimistic disposition by using the maximum reward observed, and then transform into average reward learners. We show both analytically and empirically that this technique can improve the convergence and stability of the learning, and is able to deal robustly with overgeneralization, miscoordination, and high degree of stochasticity in the reward and transition functions. Our method outperforms state-of-the-art multi-agent learning algorithms across a spectrum of stochastic and partially observable games, while requiring little parameter tuning.

Keywords: Multi-agent reinforcement learning · Markov games · Independent learners · Distributed Q-learning

1 Introduction

Many real-world problems involve environments that contain a group of independently learning agents working towards a common goal. Examples include air traffic control [1], adaptive load-balancing of parallel applications [14], and multi-robot systems [19]. Reinforcement learning (RL) [13] has been a popular approach to solving multi-agent learning problems, since it does not require any a-priori knowledge about the dynamics of the environment, which can be stochastic and partially observable.

Two main multi-agent reinforcement learning (MARL) approaches have been proposed for handling multi-agent domains [3]. In joint-action learners (JALs), every agent is aware of the actions of the other agents. Thus, the state and action spaces of all agents can be merged together, defining a sort of super-agent in the

© Springer Nature Switzerland AG 2020
M. E. Taylor et al. (Eds.): DAI 2020, LNAI 12547, pp. 69–84, 2020.
https://doi.org/10.1007/978-3-030-64096-5_6

joint space. Any single-agent RL algorithm can be used to learn the optimal joint policy for such super agent. On the other hand, independent learners (ILs) do not know the actions the other agents are performing, thus each agent has to learn and act independently. Independent learning is an appealing approach, since it does not have the scalability problem with increasing the number of agents, and each agent only needs its local history of observations during the training and the inference time.

However, independent learners must overcome a number of pathologies in order to learn optimal joint policies [4,8]. These include credit assignment, the moving target problem, stochasticity, relative overgeneralization, miscoordination, and the alter-exploration problem. Due to these pathologies, reinforcement learning algorithms that converge in a single-agent setting often fail in cooperative multi-agent systems with independent learning agents.

Various methods have been proposed in the MARL literature to help ILs cope with the outlined pathologies. However, addressing one pathology often leaves agents vulnerable towards others. Distributed Q-learning [5] is an optimistic learning approach, which only increases the Q-values and ignores negative updates. It solves the relative overgeneralization and miscoordination problems, but is vulnerable to games with stochastic rewards or transitions. Other approaches, such as hysteretic Q-learning [8] and lenient learning [12] are based on varying the "degree of optimism" of agents. Hysteretic Q-learning uses a variable learning rate to minimize the influence of poor actions of other agents on the agent's own learning, while lenient Q-learning initially ignore rewards that lead to negative updates, until the state-action pair has been encountered frequently enough (see Sect. 5 for a more detailed description).

All previous methods use the latest joint reward signal received for each state-action pair to update the Q function. Since the joint reward depends both on the behavior of the other agents (which initially are mostly exploring the environment), and the noise in the environment, updating the Q values and the corresponding policy using only the latest reward signal may destabilize the learning.

Instead, we propose to keep track of the rewards received for each state-action pair, and use the reward statistics for the policy updates. Similar to the hysteretic and lenient approaches, we use a hybrid approach for the learning: initially, the agents adopt an optimistic approach by using the maximum reward observed for the updates, which helps minimize the effect of shadowed equilibria. After a predefined period of time, the agents transform into average reward learners in order to handle stochasticity. We call this approach *Hybrid Q-learning*.

We first analyze the Hybrid Q-learning algorithm, and show that it is able to handle each of the outlined MARL pathologies. We then compare it against baseline algorithms from the literature, including traditional Q-learning, Distributed Q-learning [5], Hysteretic Q-learning [8], and a recent version of Lenient Q-Learning called LMRL2 [18]. For the comparison we use a diverse array of situations, including: stochastic and repeated games, deterministic and stochastic rewards, stochastic state transition functions, partial observability, miscoordination, and relative overgeneralization. Our method outperforms the baseline algorithms on all tested problems.

2 Theoretical Framework

2.1 Markov Games

Markov games (also called Stochastic games) [7] are the foundation for much of the research in multi-agent reinforcement learning. Markov games are a superset of Markov decision processes (MDPs) and matrix games, including both multiple agents and multiple states.

Formally, a Markov game consists of a tuple $\langle N, S, A, T, R \rangle$ where:

- N is a finite set of agents, $n = |N|$
- S is a set of states
- $A = A_1 \times ... \times A_n$ is the set of joint actions.
- $T : S \times A \times S \rightarrow [0, 1]$ is the transition probability function, with $T(s, a, s') = P(s'|s, a)$ being the probability of transitioning to state s' after taking joint action a in state s.
- $R : S \times A \rightarrow \mathbb{R}^n$ is the joint reward function, with $R_i(s, a)$ being the reward of agent i in state s after taking the joint action a.

In a Markov game, each agent is independently choosing actions and receiving rewards, while the state transitions with respect to the full joint-action of all agents. If all agents receive the same rewards ($R_1 = ... = R_n = R$) the Markov game is *fully-cooperative*.

2.2 Policies and Nash Equilibria

For agent i, the policy π_i represents a mapping from a state to a probability distribution over actions: $\pi_i : S \times A \rightarrow [0, 1]$. A solution to a stochastic game is a *joint policy* π—a set of policies, one for each agent. Joint policies excluding agent i are defined as π_{-i}. The notation $\langle \pi_i, \pi_{-i} \rangle$ refers to a joint policy with agent i following π_i while the remaining agents follow π_{-i}.

Given a joint policy π, the value (or gain) of state s for agent i is defined as the expected sum of discounted future rewards, when all agents follow π from state s:

$$V^{\pi_i}(s) = \mathbb{E}_\pi \left[\sum_{k=0}^{\infty} \gamma^k r_{i,t+k} \middle| s_t = s \right] \tag{1}$$

where $r_{i,t}$ is the reward received by agent i at time step t, and $\gamma \in (0, 1]$ is a discount factor.

From a game theoretic point of view, two common concepts of equilibrium are used to define solutions in Markov games. A joint policy π^* is a *Nash equilibrium*

[10], if and only if no agent can improve its gain by unilaterally deviating from π^*, i.e., for agent i

$$\forall \pi_i, \forall s \in S, V^{\langle \pi_i^*, \pi_{-i}^* \rangle}(s) \geq V^{\langle \pi_i, \pi_{-i}^* \rangle}(s) \tag{2}$$

In contrast, a Pareto-optimal equilibrium is a joint policy $\hat{\pi}$ from which no agent can improve its gain without decreasing the expected gain of any other agent. In cooperative games, every Parto-optimal equilibrium is also a Nash equilibrium. Thus, the objective in cooperative games is to find the Paerto-optimal Nash equilibria, also called *optimal equilibria*. Since there can be several optimal equilibria, some coordination mechanisms are necessary.

2.3   Q-Learning

Q-learning [17] is one of the most popular single-agent reinforcement learning algorithms, due to its simplicity and robustness. That is why it was one of the first algorithms to be applied to multi-agent settings.

Let $Q_i(s, a_i)$ be the value of the state-action pair (s, a_i) for agent i, where a_i is the agent's chosen action.[1] The update equation for agent i is:

$$Q_i(s, a_i) \leftarrow (1 - \alpha)Q_i(s, a_i) + \alpha \left[r + \gamma \max_{a \in A_i} Q_i(s', a) \right] \tag{3}$$

where s' is the new state, $r = R(s, \langle a_i, a_{-i} \rangle)$ is the global reward received, $\alpha \in [0, 1]$ is the learning rate, and $\gamma \in [0, 1]$ is the discount factor.

3   Hybrid Q-Learning

In this section we present our Hybrid Q-learning approach for solving cooperative stochastic games (shown in Algorithm 1). The algorithm combines between two types of independent learners, that attempt to estimate the quality of an action a, when paired with the actions A' of the other agents [18]:

- *Maximum-based learners* estimate the quality of a based on the maximum reward observed:
$$\text{quality}(a) = \max_{a' \in A'} R_i(a, a')$$

- *Average-based learners* estimate the quality of a based on the average reward:
$$\text{quality}(a) = \frac{\sum_{a' \in A'} R_i(a, a')}{|A'|}$$

---

[1] We sometimes omit the subscript i, when it is clear that we are referring to a specific agent.

Our agents start as maximum-based learners, by using the maximum reward obtained in each state-action pair for the Q updates. This allows the agents to be initially forgiving towards suboptimal actions by teammates, which are mostly doing exploration in that phase. It also prevents premature convergence of the agents upon a sub-optimal joint policy (see Sect. 4 for a formal analysis).

After the agents have gained enough information about the potential rewards that may be obtained in each state-action pair, they transition into using the average rewards for the Q updates. This allows the agents to learn whether the variation in the observed rewards is caused due to the noise in the environment or to the behaviors of the other agents, thus making them more robust against stochastic rewards and transitions.

Note that the average rewards are being computed from the beginning of the game, thus all the knowledge the agents have accumulated during the exploration phase is being utilized. This is in contrast to other types of learners which initially ignore rewards that lead to negative updates.

A parameter denoted by ρ ($0 \leq \rho \leq 1$) controls the number of "optimistic" steps, during which the maximum reward is used for the Q updates. After ρT number of steps is reached (T is the time horizon of the game), the agents move to using the average rewards for the updates. In our experiments ρ was set to be within the range of $[0.1\text{–}0.2]$.

In order to keep track of the reward estimates, each agent maintains four tables:

1. $Q(s,a)$, a table of Q-values per state and action
2. $R_{\max}(s,a)$, a table of maximum rewards per state and action
3. $\overline{R}(s,a)$, a table of average rewards per state and action
4. $N(s,a)$, the number of visits to each state and action

To compute the average rewards in a computationally efficient manner, we make incremental updates to the average rewards received at a state-action pair (s,a) using the following equation:

$$\overline{R}_{n+1}(s,a) = \overline{R}_n(s,a) + \frac{1}{n+1}\left(r_{n+1} - \overline{R}_n(s,a)\right) \tag{4}$$

where r_{n+1} is the reward received after the $(n+1)$th visit to (s,a).

$Q(s,a)$ is updated in Q-learning style to incorporate both the immediate reward and the expected future utility. However, instead of using the current reward signal, which may be unreliable due to noise from the environment and the other agents' actions, we use either the maximum reward $R_{\max}(s,a)$ or the average reward $\overline{R}(s,a)$, depending on the current stage of the learner.

The extension of Algorithm 1 to partially observable environments is pretty straightforward. In the Q and R tables, instead of using the state variable we use the agent's observation history, e.g., $Q(s,a)$ changes into $Q(h,a)$.

Algorithm 1. Hybrid Q-Learning for agent i

Input: Max steps T, ratio of optimistic updates ρ, learning rate α, discount factor γ, exploration ratio ϵ

for $s \in S$ and $a \in A_i$ **do**
    $Q(s, a) \leftarrow 0$
    $R_{\max}(s, a) \leftarrow -\infty$
    $\overline{R}(s, a) \leftarrow 0$
    $N(s, a) \leftarrow 0$

$s \leftarrow$ initial state

for $t \leftarrow 1$ to T **do**
    From s select a using an ϵ-greedy selection rule
    Apply a and observe reward r and next state s'
    $R_{\max}(s, a) \leftarrow \max(R_{\max}(s, a), r)$
    $N(s, a) \leftarrow N(s, a) + 1$
    $\overline{R}(s, a) \leftarrow \overline{R}(s, a) + \frac{1}{N(s,a)}(r - \overline{R}(s, a))$
    if $t < \rho$ **then**
        $R \leftarrow R_{\max}(s, a)$
    else
        $R \leftarrow \overline{R}(s, a)$
    $Q(s, a) \leftarrow (1 - \alpha)Q(s, a) + \alpha[R + \gamma \max_{a' \in A_i} Q(s', a')]$
    $s \leftarrow s'$

---

4   Pathologies in Multi-Agent RL

In this section we discuss typical pathologies found in multi-agent learning problems, and describe how our approach tackles each one of them.

4.1   Relative Overgeneralization

Relative overgeneralization is a type of equilibrium shadowing, occurring when a sub-optimal Nash equilibrium yields a higher payoff on average, when an agent's selected action is paired with arbitrary actions chosen by the other agents. As a result of the shadowed equilibrium, average-based learners converge upon the sub-optimal Nash equilibrium, that is pareto-dominated by at least one other Nash equilibrium.

For example, in the Climb game (Table 1), the joint action $\langle a, a \rangle$ defines a Pareto-optimal Nash equilibrium. However, if the agents sampled their actions randomly and based their decisions on the average reward, action a would have the worst score (-6.333), action c would have the best score (1.667), and b would be in the middle (-5.667). Thus, average-based learners would tend to converge to the joint action $\langle c, c \rangle$, i.e., the Nash equilibrium is shadowed by the joint action $\langle c, c \rangle$.

On the other hand, maximum-based learners would have evaluated action a as the best action, c as the second best, and b as the worst. Hence, these agents would tend to converge to the proper equilibrium $\langle a, a \rangle$.

In our hybrid approach, the agents start as maximum-based learners. Thus, given enough time for initial exploration, once both agents choose action a, it will become their preferred action. Transitioning into average-based learners will not change their policy, since after both agents' policies have converged to the joint action $\langle a, a \rangle$, the average reward for taking action a will be significantly higher than taking any other action.

Table 1. The Climb game

	Agent 2		
	a	b	c
a	11	-30	0
b	-30	7	6
c	0	0	5

(Agent 1 labels rows a, b, c)

4.2   The Stochasticity Problem

In stochastic games, the difficulty for independent learners is to distinguish whether the variation in the observed rewards is due to the noise in the environmnet or to the other agent's behaviors.

For example, in the partially stochastic Climb game (Table 2a) the joint action $\langle b, b \rangle$ yields stochastic rewards of 14 and 0 with 50% probability. An IL must distinguish if distinct rewards received for action b are due to various behaviors of the other agent or to stochastic rewards. In this case, maximum-based learners are drawn towards $\langle b, b \rangle$, despite each agent only receiving a reward of 7 on average. On the other hand, average-based learners learn that the average reward of the joint action $\langle b, b \rangle$ is 7, and thus that the action b is sub-optimal.

Table 2. Stochastic variations of the climb game. The probability of each reward is 50%.

	Agent 2		
	a	b	c
a	11	-30	0
b	-30	14|0	6
c	0	0	5

(a) Climb-PS

	Agent 2		
	a	b	c
a	10|12	5|(−65)	8|(−8)
b	5|(−65)	14|0	12|0
c	8|(−8)	12|0	10|0

(b) Climb-FS

Using our hybrid learning approach can solve this problem. In the optimistic training phase, each agent will learn the maximum attainable reward for each action, i.e., 11 for a, 14 for b, and 5 for c. Then, after the transition into using the average rewards, the reward for action b will start to drop off towards 7, thus the agents will switch to the second best action a and stick to it (since its average reward will be the same as the maximum).

Stochastic transitions present an additional hurdle. As was the case for stochastic rewards, stochastic transitions are largely overcome by the agents' gradual shift from being maximum-based to being average-based.

4.3   Miscoordination

Miscoordination occurs when there are multiple Pareto-optimal equilibria, and they are offset in such a way that agents, each selecting what appears to be an optimal action, end up with poor joint actions. For example, in the Panelty game (Table 3), both actions a and c are reasonable for each agent, but if one agent chooses a and the other one chooses c, they receive a low reward.

Table 3. The penalty game

		Agent 2		
		a	b	c
Agent 1	a	10	0	-10
	b	0	2	0
	c	-10	0	10

With our Hybrid-Q approach, if Agent 1 keeps choosing action a and Agent 2 keeps choosing action c, when they turn into average-reward learners, they will both receive an average reward of 5. At that point, when one of the agents tries the other action (as exploration continues with a small probability), both agents will start receiving an average reward that is greater than 5, which will keep them selecting the same action.

4.4   The Alter-Exploration Problem

In multi-agent settings, the exploration of one agent induces noise in received rewards of other agents and can destabilize their learned policies. This alter-exploration problem amplifies the issue of shadowed equilibria, as a high global exploration can result in agents converging upon a sub-optimal joint policy [9].

Our approach handles the alter-exploration issue by adopting a maximum-based learning approach during the exploration phase, i.e., it allows the learning of individual policies while other agents are exploring. By the time the agents have gathered enough data about the rewards, they can change into average-based learners without having the risk of being trapped in a sub-optimal policy due to the other agents' exploring actions.

5   Independent Learner Baselines

We compare our Hybrid Q-learning against four other independent learner algorithms in several cooperative test problems. We briefly discuss these algorithms below.

5.1   Independent Q-Learning

In a MARL context, standard Q-learning is also known as Independent Q-learning (IQL) or Decentralized Q-learning. In IQL, each agent runs an independent Q-learning algorithm with its own Q value estimates, and there is no mechanism for addressing coordination problems. This algorithm has a low robustness face to exploration, since explorative actions by other agents can cause the agent's learned policy to oscillate. Nevertheless, IQL has been successfully applied in some domains [1,6,16].

5.2   Distributed Q-Learning

In Distributed Q-learning [5], the agents optimistically assume that all the other agents will act to maximize their reward. Thus, they update the Q-value of an action only if it is greater than the previous one. This optimistic update avoids sub-optimal policies caused by shadowed equilibria.

To deal with miscoordination, an additional procedure is used to solve the Pareto-optimal equilibria selection problem: the current policy π_i is only updated when the Q-value of the policy's chosen action is no longer the highest such value. The idea is to cause agents to lock onto the earliest-discovered good action (and Nash equilibrium) even when other equivalent equilibria exist.

Distributed Q-learning was proved to find optimal policies in deterministic cooperative Markov games. However, this approach generally does not converge in stochastic environments. In such environments, optimistic learners ignore penalties in the update which are caused by noise in the environment, thus they overestimate the real Q values.

5.3   Hysteretic Q-Learning

Hysteretic Q-learning is an optimistic MARL algorithm introduced to address maximum based learner's vulnerability towards stochasticity [8]. The idea is that agents on one hand should not be blind to penalties, and on the other hand they must be optimistic to minimize the effect of shadowed equilibria.

Hysteretic Q-learning uses two learning rates α and β for positive and negative updates of the Q values, where $\beta < \alpha$. The smaller β learning rate reduces the impact of negative Q-value updates.

However, hysteretic Q-learners still have a tendency to converge to suboptimal policies when receiving misleading stochastic rewards. [18]

5.4   LMRL2

LMRL2 (Lenient Multiagent Reinforcement Learning 2) [18] is an extension of the lenient learning algorithm, LMRL [12], for stochastic games. LMRL2 is a modified version of Q-learning which maintains per-action temperatures, that are slowly decreased throughout the learning process. An action's temperature affects two things. First, it affects the lenience of the algorithm: high temperatures cause the algorithm to be lenient towards negative policy updates, and so only mix rewards into Q-values if they are better than the current Q-value. With a low temperature, LMRL2 mixes all rewards into the Q value.

Second, the temperature affects the degree of exploration: with high temperatures, action selection is largely random, and with low temperatures, action selection is greedily based on the actions with the highest Q-values. However, when the average temperature drops below a certain minimum temperature, then LMRL2's action selection will suddenly become purely greedy.

5.5   Parameters

All the mentioned techniques have a variety of tunable parameters. For the benchmarks used in our experiments, we have applied the same parameter settings as in [18] (also shown in Table 4). Note that some of these parameters are unique to specific algorithms, e.g., the temperature parameters are relevant only for lenient learning. Furthermore, some parameter values had to be tuned for specific problems in order to obtain the best policy.

Table 4. Default parameter settings

Discount factor	γ	0.9
Learning rate	α	0.1
Exploration ratio	ϵ	0.1
Hysteretic learning rate	β	0.02
Temprature diffusion coefficient	τ	0.1
Temprature decay coefficient	δ	0.995
Minimum temprature	MinTemp	50.0
Maximum temprature	MaxTemp	2.0
Action selection moderation factor	ω	0.1
Leniency moderation factor	θ	1

6   Experiments

We compared our method against all the aforementioned algorithms using four types of test problems with various degrees of stochasticity and partial observability. Each of these problems exhibits some degree of the multi-agent pathologies discussed in Sect. 4.

6.1 Climb Games

First, we apply our algorithm to the three versions of the Climb game presented in Tables 1 and 2. These games have received considerable attention from the community [3,5,9,11,18], as they remain challenging for state-of-the-art algorithms despite their apparent simplicity.

The partially stochastic and fully stochastic versions, here designated Climb-PS and Climb-FS respectively, define the reward for a joint action as one of two different values with certain probabilities. However, the expected reward is always the same as in the original Climb game. Stochasticity greatly complicates the problem: past learners have failed to coordinate in the fully-stochastic Climb game [9,18].

For the baseline algorithms, some of the parameters were tuned according to the values suggested in [18]. For the Hybrid-Q algorithm, we have used the same parameter values for all three versions of the game: $\rho = 0.1$, $\alpha = 0.1$, $\gamma = 0.9$, and $\epsilon = 0.01$. We have found out that higher exploration rates destabilize the learned policy when the agents transition into average-based learners (interestingly, small exploration rates have also been used in [18] for the baseline algorithms).

We have tested each algorithm on 10,000 trials of the game with different random seeds. A trial consists of 10,000 repetitions of the game ($T = 10000$). At the end of each trial, we determine if the greedy joint action is the optimal one. Table 5 shows the percentage of trials converging to the optimal joint action according to the algorithm and the game.

Table 5. Percentage of trials that converged to the optimal joint action in the Climb games (averaged over 10,000 trials). Boldface values indicate the highest-performing methods for a given problem.

	Climb	Climb-PS	Climb-FS
Hybrid Q	**100%**	**99.65%**	**99.45%**
IQL	58.80%	58.93%	50.34%
Distributed Q	**100%**	28.83%	38.11%
Hysteretic Q	**100%**	93.92%	94.98%
LMRL2	**100%**	71.30%	35.11%

We can notice that our Hybrid-Q approach outperforms all baseline algorithms in the stochastic versions of the game, and on par with the best algorithms in the deterministic version.

6.2 Heaven and Hell Game

The Heaven and Hell game [18] is a deterministic game with four states and two actions for each player (Table 6). This game has a high-reward state ("heaven"), a low-reward state ("hell"), and two medium ("purgatory") states of different levels of reward. Choice of a high-reward action will deceptively transition to a lower-reward future state, and the converse is also true.

Table 6. Heaven and Hell game. The optimal joint action in each state is indicated with boldface.

S	A_1	A_2	Reward	S'
s_1	a	a	20	s_2
s_1	**any**	**other**	10	s_1
s_1	b	b	15	s_3
s_2	a	a	0	s_4
s_2	**any**	**other**	−10	s_1
s_2	b	b	−5	s_2
s_3	a	a	10	s_4
s_3	**any**	**other**	0	s_1
s_3	b	b	5	s_2
s_4	a	a	−10	s_4
s_4	**any**	**other**	−20	s_3
s_4	b	b	−15	s_2

In this game, it is easy for all algorithms to find that the best state for the agents to be in is s_1, thus any joint action that keeps the agents in that state (either $\langle a, b \rangle$ or $\langle b, a \rangle$) is optimal. Therefore, we have also checked which of the algorithms has found the more general *complete policy*, meaning a policy that determines the correct joint action for every state, even ones which would never be visited if it followed a correct policy.

As in the Climb games, each trial consisted of 10,000 repetitions of the game. This time we have conducted 1,000 trials for each algorithm. Table 7 shows the percentage of trials converging to a complete and optimal joint policy according to the algorithm.

Table 7. Percentage of trials that converged to the complete and optimal joint policy in the heaven and hell problem (averaged over 1,000 trials)

Hybrid Q	IQL	Distributed Q	Hysteretic Q	LMRL2
71.0%	17.2%	23.1%	56.8%	63.7%

We can see that the success ratio in finding complete policies is lower than what was achieved in the Climb games, since in this game a complete policy is not needed for maximizing the total accumulated reward.

6.3 Common Interest Game

In contrast to the previous games, the Common Interest game [15] is a stochastic game with two states and stochastic transitions between these states. Table 8 shows the rewards and the transition probabilities for each state and joint action. This game poses a miscoordination challenge, as there are two Pareto-optimal equilibria in the game. In the first, the agents play joint action (a, b) in state 1 and again (a, b) in state 2, while in the second they play (b, a) in state 1 and (a, b) in state 2.

Table 8. Common interest game

S	A_1	A_2	Reward	S
s_1	a	a	0.5	$s_1(10\%)$, $s_2(90\%)$
s_1	a	b	0.6	$s_1(10\%)$, $s_2(90\%)$
s_1	b	a	0.6	$s_1(10\%)$, $s_2(90\%)$
s_1	b	b	0.7	$s_1 (90\%)$, $s_2 (10\%)$
s_2	a	a	0	$s_1(10\%)$, $s_2(90\%)$
s_2	a	b	1.0	$s_1(10\%)$, $s_2(90\%)$
s_2	b	a	0.5	$s_1(10\%)$, $s_2(90\%)$
s_2	b	b	0	$s_1(90\%)$, $s_2(10\%)$

Similarly to the Climb games, we have executed the Common Interest game for 10,000 time steps, and tested each algorithm on 10,000 instances of the game with different random seeds. At the end of each game, we checked if the final joint policy is one of the two Pareto-optimal equilibria. Table 9 shows the percentage of trials converging to an optimal joint policy according to the algorithm. As can be seen, the Hybrid-Q algorithm has achieved the highest percentage of optimal solutions.

Table 9. Percentage of trials that converged to the optimal joint action in the Common Interest problem (averaged over 10,000 trials)

Hybrid Q	IQL	Distributed Q	Hysteretic Q	LMRL2
99.91%	91.70%	30.58%	59.90%	98.79%

6.4 Meeting in a Grid

In our last experiment, we have tested our algorithm on a partially observable and stochastic environment. We used a variation of the meeting-in-a-grid Dec-POMDP benchmark [2]. The environment consists of a static target and two agents in a 8×8 toroidal grid world (Fig. 1). The agents get a reward of 1 for simultaneously landing on the target location, otherwise a reward of -0.01 (which encourages the agents to reach the target as quickly as possible). Episodes terminate after 40 transitions or upon successful capturing of the target. Each agent observes its own location and the location of the target, but does not know where the other agent is located.

Each agent can execute one of the five possible actions: move North, South, East, West, or stand still. Actions result in stochastic transitions: each action succeeds with probability $1 - p$, but with probability p, the agent moves at right angles to the intended direction. For example, if $p = 0.2$, and the agent is located at cell (2, 2) and chooses to move North, then with probability 0.8 it will move to cell (2, 1), but with probability 0.1 it will move left to cell (1, 2), and with probability 0.1 it will move right to cell (3, 2).

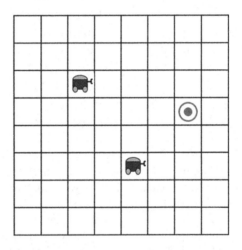

Fig. 1. The meeting-in-a-grid environment.

We ran 50 experiments with different random locations of the agents and the target, each consisting of 1,000 episodes. For each experiment we measure the total return over all the episodes. Therefore, the maximum score that can be obtained in each experiment is 1,000, which occurs when the agents learn to capture the target on the first episode, and keep capturing it in all subsequent episodes.

We have varied the action noise probability p from 0.1 to 0.5 by intervals of 0.05. Figure 2 shows the average return achieved by the various algorithms for the different values of p. As the action noise level increases, the gap between the Hybrid-Q method and the other algorithms widens, which indicates that our approach can better handle higher levels of stochasticity in the environment.

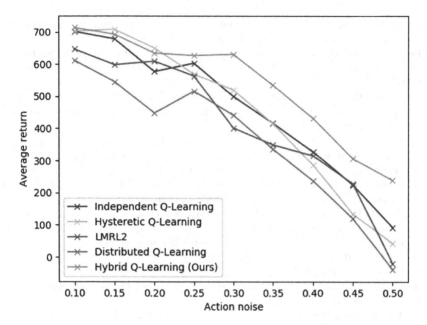

Fig. 2. Average return in meeting-in-a-grid task for various action noise levels.

7 Conclusions

We presented a new hybrid approach for multi-agent reinforcement learning, which is based on agents transitioning from being maximum-based learners into average-based learners. We have shown that our approach consistently converges towards the optimal joint policy in several benchmarks from the multi-agent learning literature.

Despite the simplicity of our approach, it can serve as a basis for a general framework that could be easily extended to other settings as well, e.g., integrating it with deep reinforcement learning, or using smooth transitions from maximum-based learning to average-based learning.

Acknowledgments. This work is funded by the U.S. Air Force Research Laboratory (AFRL), BAA Number: FA8750-18-S-7007, and NSF grant no. 1816382.

References

1. Agogino, A.K., Tumer, K.: A multiagent approach to managing air traffic flow. Auton. Agent. Multi-Agent Syst. **24**(1), 1–25 (2012). https://doi.org/10.1007/s10458-010-9142-5
2. Amato, C., Dibangoye, J.S., Zilberstein, S.: Incremental policy generation for finite-horizon DEC-POMDPs. In: International Conference on Automated Planning and Scheduling (ICAPS) (2009)

3. Claus, C., Boutilier, C.: The dynamics of reinforcement learning in cooperative multiagent systems. AAAI/IAAI **1998**, 746–752 (1998)
4. Fulda, N., Ventura, D.: Predicting and preventing coordination problems in cooperative q-learning systems. Int. Joint Conf. Artif. Intell. (IJCAI). **2007**, 780–785 (2007)
5. Lauer, M., Riedmiller, M.: An algorithm for distributed reinforcement learning in cooperative multi-agent systems. In: International Conference on Machine Learning (ICML) (2000)
6. Leibo, J.Z., Zambaldi, V., Lanctot, M., Marecki, J., Graepel, T.: Multi-agent reinforcement learning in sequential social dilemmas. In: International Conference on Autonomous Agents and Multiagent Systems (AAMAS), pp. 464–473 (2017)
7. Littman, M.L.: Markov games as a framework for multi-agent reinforcement learning. In: International Conference on Machine Learning (ICML), pp. 157–163 (1994)
8. Matignon, L., Laurent, G.J., Le Fort-Piat, N.: Hysteretic q-learning: an algorithm for decentralized reinforcement learning in cooperative multi-agent teams. In: IEEE/RSJ International Conference on Intelligent Robots and Systems (IROS), pp. 64–69 (2007)
9. Matignon, L., Laurent, G.J., Le Fort-Piat, N.: Independent reinforcement learners in cooperative markov games: a survey regarding coordination problems. Knowl. Eng. Rev. **27**(1), 1–31 (2012)
10. Nash, J.F.: Equilibrium points in n-person games. Proc. Natl. Acad. Sci. U.S.A. **36**(1), 48–49 (1950)
11. Palmer, G., Savani, R., Tuyls, K.: Negative update intervals in deep multi-agent reinforcement learning. In: International Conference on Autonomous Agents and MultiAgent Systems (AAMAS), pp. 43–51 (2019)
12. Panait, L., Sullivan, K., Luke, S.: Lenient learners in cooperative multiagent systems. In: International Conference on Autonomous Agents and Multiagent Systems (AAMAS), pp. 801–803 (2006)
13. Sutton, R.S., Barto, A.G.: Reinforcement Learning: An Introduction. MIT Press, Cambridge (2018)
14. Verbeeck, K., Nowé, A., Parent, J., Tuyls, K.: Exploring selfish reinforcement learning in repeated games with stochastic rewards. Auton. Agent. Multi-Agent Syst. **14**(3), 239–269 (2007)
15. Vrancx, P., Tuyls, K., Westra, R.: Switching dynamics of multi-agent learning. Int. Conf. Auton. Agent. Multiagent Syst. (AAMAS) **1**, 307–313 (2008)
16. Wang, Y., De Silva, C.W.: Multi-robot box-pushing: single-agent q-learning vs. team q-learning. In: IEEE/RSJ International Conference on Intelligent Robots and Systems (IROS), pp. 3694–3699 (2006)
17. Watkins, C.J., Dayan, P.: Q-learning. Mach. Learn. **8**(3–4), 279–292 (1992)
18. Wei, E., Luke, S.: Lenient learning in independent-learner stochastic cooperative games. J. Mach. Learn. Res. **17**(1), 2914–2955 (2016)
19. Yang, E., Gu, D.: Multiagent reinforcement learning for multi-robot systems: A survey. Technical report, Department of Computer Science, University of Essex, Technical report (2004)

Efficient Exploration by Novelty-Pursuit

Ziniu Li[1,2]($\boxtimes$) and Xiong-Hui Chen[3]($\boxtimes$)

[1] Polixir, Nanjing, China
ziniu.li@polixir.ai
[2] The Chinese University of Hong Kong, Shenzhen, Shenzhen, China
ziniuli@link.cuhk.edu.cn
[3] National Key Laboratory for Novel Software Technology, Nanjing University,
Nanjing, China
chenxh@lamda.nju.edu.cn

Abstract. Efficient exploration is essential to reinforcement learning in tasks with huge state space and long planning horizon. Recent approaches to address this issue include the intrinsically motivated goal exploration processes (IMGEP) and the maximum state entropy exploration (MSEE). In this paper, we propose a goal-selection criterion in IMGEP based on the principle of MSEE, which results in the new exploration method *novelty-pursuit*. Novelty-pursuit performs the exploration in two stages: first, it selects a seldom visited state as the target for the goal-conditioned exploration policy to reach the boundary of the explored region; then, it takes random actions to explore the non-explored region. We demonstrate the effectiveness of the proposed method in environments from simple maze environments, MuJoCo tasks, to the long-horizon video game of SuperMarioBros. Experiment results show that the proposed method outperforms the state-of-the-art approaches that use curiosity-driven exploration.

Keywords: Reinforcement learning · Markov decision process · Efficient exploration

1 Introduction

Reinforcement learning (RL) [39,40] is a learning paradigm that an agent interacts with an unknown environment to improve its performance. Since the environment transition is unknown in advance, the agent must explore (e.g., take new actions) to discover states with positive rewards. Hence, efficient exploration is important to learn a (near-) optimal policy in environments with huge state space and sparse rewards [44], where deep-sights planning behaviors are required [29]. In those cases, simple exploration strategies like ϵ-greedy are inefficient due to time-uncorrelated and uncertainty-unaware behaviors [2].

To avoid insufficient exploration, advanced curiosity-driven approaches encourage diverse actions by adding the uncertainty-based exploration bonus

---

Z. Li and X-H. Chen—The two authors contributed equally to this work.

© Springer Nature Switzerland AG 2020
M. E. Taylor et al. (Eds.): DAI 2020, LNAI 12547, pp. 85–102, 2020.
https://doi.org/10.1007/978-3-030-64096-5_7

on the environment reward [4,7,30,31,37]. In addition, recently proposed methods to tackle this issue include the intrinsically motivated goal exploration processes (IMGEP) [14], and the maximum state entropy exploration (MSEE) [18]. In particular, IMGEP was biologically-inspired to select intrinsically interesting states from the experience buffer as goals and to train a goal-conditioned (goal-input) exploration policy to accomplish the desired goals. On the other hand, MSEE aimed to search for a policy such that it maximizes the entropy of state distribution.

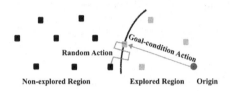

Fig. 1. Illustration for the proposed method. First, a goal-conditioned policy plans to reach the exploration boundary; then, it performs random actions to discover new states.

In this paper, we propose a goal-selection criterion for IMGEP based on the principle of MSEE, which results in the new exploration method *novelty-pursuit*. Abstractly, our method performs in two stages: first, it selects a novel state that was seldom visited as the target for the goal-conditioned exploration policy to reach the boundary of the explored region; subsequently, it takes random actions to discover new states. An illustration is given in Fig. 1. Intuitively, this process is efficient since the agent avoids meaningless exploration within the explored region. Besides, the exploration boundary will be expanded further as more and more new states are discovered. To leverage good experience explored by the goal-conditioned policy, we also train an unconditioned policy that exploits collected experience in the way of off-policy learning.

We conduct experiments on environments from simple maze environments, MuJoCo tasks, to long-horizon video games of SuperMarioBros to validate the exploration efficiency of the proposed method. In particular, we demonstrate that our method can achieve a large state distribution entropy, which implies our exploration strategy prefers to uniformly visit all states. Also, experiment results show that our method outperforms the state-of-the-art approaches that use curiosity-driven exploration.

2   Related Work

In addition to the ϵ-greedy strategy, simple strategies to remedy the issue of insufficient exploration include injecting noise on action space [23,26] or parameter space [8,10,15,24,34], and adding the policy's entropy regularization [25,36]. For the tabular Markov Decision Process, there are lots of theoretical studies utilizing upper confidence bounds to perform efficient exploration [21,22,38]. Inspired by this, many deep RL methods use curiosity-driven exploration strategies [4,7,30,31,37]. In particular, these methods additionally add the uncertainty-based exploration bonus on the environment reward to encourage diverse behaviors. Moreover, deep (temporally extended) exploration via tracking the uncertainty of value function was studied in [29]. Maximum (policy) entropy reinforcement learning, on the other hand, modifies the original objective function and encourages exploration by incorporating the policy entropy into the environment reward [17,28].

Our method is based on the framework of intrinsically motivated goal exploration processes (IMGEP) [3,14,32]. Biologically inspired, IMGEP involves the following steps: 1) selecting an intrinsically interesting state from the experience buffer as the goal; 2) exploring with a goal-conditioned policy to accomplish the target; 3) reusing experience by an exploitation policy that maximizes environment rewards. Obviously, the performance of exploitation policy heavily relies on samples collected by the goal-conditioned exploration policy so that the criterion of intrinsic interest is crucial for IMGEP. As reminiscent of IMGEP, Go-Explore [12] used the heuristics based on visitation counts and other domain knowledge to select goals. However, different from the basic framework of IMGEP, Go-Explore directly reset environments to target states, upon which it took random actions to explore. As a result, it achieved dramatic improvement in the challenging exploration task of Montezuma's Revenge. However, the requirement that the environment is resettable (or the environment transition is deterministic), together with many hand-engineering designs, clearly restricts Go-Explore's applications. The exploration scheme of our method in Fig. 1 is similar to Go-Explore, but our method does not require environments to be resettable or deterministic.

Recently, [18] introduced a new exploration objective: maximum state entropy. Since each policy induces a (stationary) state distribution, [18] provided a provably efficient reinforcement learning algorithm under the tabular MDP to search for an optimal policy such that it maximizes the state (distribution) entropy. The such-defined optimal policy is conservative to visit uniformly all states as possible in unknown environments. Inspired by this principle, we propose to select novel states as goals for the goal-conditioned exploration policy in IMGEP. Note that [18] mainly focused on the pure exploration problem while we additionally train an exploitation policy that leverages collected experience in the way of off-policy learning to maximize environment rewards.

Finally, we briefly review recent studies about how to quickly train a goal-conditioned policy since it is a core component of our method. Particularly, [35] proposed the universal value function approximator (UVFA) and trained it by bootstrapping from the Bellman equation. However, this training procedure is still inefficient because goal-conditioned rewards are often sparse (e.g. 1 for success and 0 for failure). To remedy this issue, [1] developed the hindsight experience replay (HER) that replaced the original goal with an achieved goal. As a result, the agent can receive positive rewards even though it does not accomplish the original goal. In this way, learning on hindsight goals may help generalize for unaccomplished goals. Moreover, [13] used a generator neural network to adaptively produce artificial feasible goals to accommodate different learning levels, which servers as an implicit curriculum learning.

3 Background

In the standard reinforcement learning (RL) framework [39,40], a learning agent interacts with an Markov Decision Process (MDP) to improve its performance via maximizing cumulative rewards. The sequential decision process is characterized as follows: at each timestep t, the agent receives a state s_t from the environment and selects an action a_t from its policy $\pi(s,a) = \Pr\{a = a_t|s = s_t\}$; this decision is sent back to the environment, and the environment gives a reward signal $r(s_t, a_t)$ and transits to the next state s_{t+1} based on the state transition probability $p^a_{ss'} = \Pr\{s' = s_{t+1}|s = s_t, a = a_t\}$. This process repeats until the agent encounters a terminal state after which the process restarts.

The main target of reinforcement learning is to maximize the (expected) episode return $\mathbb{E}[\sum_{t=0}^{\infty} \gamma^t r(s_t, a_t)]$, where $\gamma \in (0,1)$ is a discount factor that balances the importance of future rewards and the expectation is taken over the stochastic process induced by the environment transition and the action selection. Since the environment transition (and possibly the reward function) is unknown in advance, the agent needs exploration to discover valuable states with positive rewards. Without sufficient exploration, the agent will be stuck into the local optimum by only learning from sub-optimal experience [44].

4 Method

As demonstrated in Fig. 1, our exploration method called *novelty-pursuit* runs in two stages: first, it selects a novel state as the goal for the goal-conditioned exploration policy to reach the boundary of the explored region; then, it takes random actions to explore the non-explored region. Again, newly observed states will be set as the desired goals in the next round. As this process repeats, the exploration boundary will be expanded further and the whole state space will probably be explored. To leverage good experience explored by the goal-conditioned policy, we also train an unconditioned policy that exploits collected experience in the way of off-policy learning. We outline the proposed approach in Algorithm 1 (for simplicity, the procedure of training the exploitation policy is omitted).

In the following parts, we focus on the detailed implementation of the proposed method. Firstly, goal-selection in complicated tasks (e.g., tasks with high-dimensional visual inputs) is given in Sect. 4.1. Then, we introduce the accelerating training techniques for goal-conditioned exploration policy in Sect. 4.2. Finally, we discuss how to effectively distill a good exploitation policy from collected experience in Sect. 4.3.

Algorithm 1. Exploration by novelty-pursuit

Input: predictor network update interval K; goal-conditioned policy update interval M; mini-batch size of samples for updating goal-conditioned policy N.

Initialize parameter θ for goal-conditioned exploration policy $\pi_g(s, g, a; \theta)$.

Initialize parameter ω_t for target network $f(\cdot; \omega_t)$, and ω_p for predictor network $\hat{f}(\cdot; \omega_p)$.

Initialize an empty experience replay buffer $\mathcal{B}$, and a priority queue Q with randomly collected states.

for each iteration **do**

    Reset the environment and get the initial state s_0;

    Choose a goal g from priority queue Q, and set $goal\_success = False$;

    for each timestep t **do**

        # Interact with the environment

        if $goal\_success == True$ **then**

            Choose an random action a_t;

        else

            Choose an action a_t according to $\pi_g(s_t, g, a_t; \theta)$;

        end if

        Send a_t to the environment, get reward r_t and the next state s_{t+1}, and update $goal\_success$;

        # Store new states and update the predictor network

        if $t \% K == 0$ **then**

            Store samples $\{s_k, g, a_k, r_k\}_{k=t-K}^{t}$ into replay buffer $\mathcal{B}$;

            Calculate prediction errors for $\{s_k\}_{k=t-K}^{t}$ and put these states into priority queue Q;

            Update predictor network $\hat{f}(\cdot; \omega_p)$ using $\{s_k\}_{k=t-K}^{t}$;

        end if

        # Update the goal-conditioned policy

        if $t \% M == 0$ **then**

            Update π_g with $\{s_k, g_k, a_k, r'_k\}_{k=1}^{N}$ sampled from $\mathcal{B}$;

        end if

    end for

end for

4.1 Selecting Goals from the Experience Buffer

As mentioned previously, our method is inspired by the principle of maximum state entropy exploration [18] to select novel states with the least visitation counts from the experience buffer as the targets. In this way, the goal-conditioned policy will spend more steps to visit the seldom visited states. As a result, the exploration policy prefers to uniformly visit all states and therefore leads to an increase in the state distribution entropy.

Unfortunately, computing visitation counts for tasks with high-dimensional visual inputs is intractable due to the curse of dimensionality. Therefore, we cannot directly apply the above goal-selection. However, it is still possible to build some variables that are related to visitation counts and are easy to compute. For example, [7] showed that prediction errors on a batch of data have a strong relationship with the number of training iterations. That is, if some samples are selected multiple times, the prediction errors of such samples will be small. Thus, we can use the prediction errors to reflect the visitation counts of observed states. Other approaches like pseudo-counts [4,30] can be also applied, but we find that the mentioned method called random network distillation (RND) by [7] is easy to scale up.

Concretely, RND is consist of two randomly initialized neural networks: a fixed (non-trainable) network called target network $f(\cdot;\omega_t)$ that is parameterized by ω_t, and a trainable network called predictor network $\hat{f}(\cdot;\omega_p)$ that is parameterized by ω_p. Both two networks take a state s as input and output a vector with the same dimension. Each time a batch of data of size K feeds into the predictor network to minimize the difference between the predictor network and the target network with respect to the predictor network's parameters (see Eq. 1).

$$\min_{\omega_p} \frac{1}{K} \sum_{i=1}^{K} ||f(s_i;\omega_t) - \hat{f}(s_i;\omega_p)||^2 \tag{1}$$

In practice, we employ an online learning setting to train RND and maintain a priority queue to store novel states based on the prediction errors of RND. In particular, after the goal-conditioned policy collects a mini-batch of states, these states will be used to train the predictor network. In this way, frequently visited states will have small prediction errors while the prediction errors for seldom visited states will be large. Also, states with large prediction errors will be stored into the priority queue and the state with the least prediction error will be removed out of the priority queue if full. This process repeats and for simplicity, no past data will be reused to train the predictor network. Based on this scheme, each iteration a novel state will be selected[1] from the priority queue as the desired goal for the goal-conditioned policy. After achieving the goal, the exploration policy will perform random actions to discover new states. Intuitively, such defined exploration behaviors will try to uniformly visit all states, which will even the state distribution and lead to an increase in state distribution entropy. This will be empirically verified in Sect. 5.

4.2 Training Goal-Conditioned Policy Efficiently

Ideally, each time we sample a novel state from the experience buffer and directly set it as the input for the goal-conditioned policy π_g. However, this processing is not friendly since the size of policy inputs is doubled and the representation of inputs may be redundant. Following prior studies about multi-goal reinforcement learning [1,33], we manually extract useful information from the state space as the input of π_g. For instance, we extract the agent position information (i.e., coordinates) from raw states, which provides a good representation of the desired goal.

---

[1] We sample goals from a distribution (e.g., softmax distribution) based on their prediction errors rather than in a greedy way.

To assign rewards for goal-conditioned policy π_g, we need to judge whether it achieves the desired goal. Again, we use the same technique to extract a representation called achieved goal ag from the observed state [1,33]. Let us denote the desired goal as g, we conclude the desired goal is accomplished if $d(ag_t, g)$ is small than a certain threshold ϵ, where ag_t is the achieved goal at timestep t and d is some distance measure (e.g., ℓ_2-norm). As a result, an ordinary method to compute rewards for the goal-conditioned policy is (note that the desired goal g does not change during an episode):

$$r'(ag_t, g) = \begin{cases} 1 & \text{if } d(ag_t, g) < \epsilon \\ 0 & \text{otherwise} \end{cases} \tag{2}$$

However, the training of goal-conditioned policy is slow with this sparse reward function. An alternative method is to use negative distance as the reward, i.e., $r' = -d(ag_t, g_t)$. However, the distance reward function may lead to unexpected behaviors [1,27]. Next, we introduce some techniques to deal with the above problems.

$$r'(ag_t, g) = d(ag_{t-1}, g) - d(ag_t, g) \tag{3}$$

Firstly, let us consider the technique of reward shaping [27], which introduces additional training rewards to guide the agent. Clearly, this operation will modify the original objective and change the optimal policy if we don't pose any restrictions. Interestingly, reward shaping is invariant to the optimal policy if the reward shaping function is a potential function [27]. Specifically, we can define the difference of two consecutive distances (between the achieved goal and the desired goal) as a reward shaping function, shown in Eq. 3. Since this function gives dense rewards, it leads to substantial acceleration in learning a near-optimal goal-conditioned policy. Consequently, the policy π_g can avoid meaningless actions within the explored region and quickly reach the exploration boundary. Verification of the optimal goal-conditioned policy is invariant under this reward shaping function is given in Appendix A.1.

Alternatively, one can use Hindsight Experience Replay (HER) [1] to train the goal-conditioned policy via replacing each episode with an achieved goal rather than one that the agent was trying to achieve. Concretely, for state s_t, we may randomly select a future state s_{t+k} from the same trajectory and replace its original target with s_{t+k}, where k is a positive index. In this way, the agent can still get positive rewards though it may not achieve the originally defined goal. But one should be careful when applying this technique since HER clearly changes the goal distribution for learning and may lead to undesired results for our setting.

4.3 Exploiting Experience Collected by Exploration Policy

In the above, we have discussed how to efficiently train a goal-conditioned policy to explore. To better leverage experience collected by the goal-conditioned policy, we need to additionally train an exploitation policy π_e that learns from goal-conditioned policy's experience buffer with environment rewards in an off-policy learning fashion.

Interestingly, it was shown that off-policy learning without interactions does not perform well on MuJoCo tasks [16]. The authors conjectured off-policy learning degenerates when the experience collected by the exploration policy is not correlated to the trajectories generated by the exploitation policy (they called this phenomenon *extrapolation error*). To remedy this issue, we add environment rewards on the goal rewards computed in the previous part. The environment rewards are scaled properly so that they do not change the original objective too much. In this way, π_e and π_g will not be too distinct. In addition, we parallelly train π_e as well as π_g and allow π_e to periodically interact with the environment to further mitigate the extrapolation error.

5   Experiment

In this section, we conduct experiments to answer the following research questions: 1) does novelty-pursuit lead to an increase in the state entropy compared with other baselines? 2) does the training technique for goal-conditioned policy improve the performance? 3) how does the performance of novelty-pursuit compare with the state-of-the-art approaches in complicated environments? We conduct experiments from simple maze environments, MuJoCo tasks, to long-horizon video games of SuperMarioBros to evaluate the proposed method. Detailed policy network architecture and hyperparameters are given in Appendix A.4 and A.5, respectively.

Here we briefly describe the environment settings (see Fig. 2 for illustrations). Details are given in Appendix A.3.

Empty Room and Four Rooms. An agent navigates in the maze of 17×17 to find the exit (the green square in Fig. 2 (a) and (b)) [9]. The agent receives a time penalty until it finds the exit and receives a positive reward. The maximal episode return for both two environments is $+1$, and the minimal episode return is -1. Note that the observation is a partially observed image of shape $(7, 7, 3)$.

FetchReach. A 7-DOF Fetch Robotics arm (simulated in the MuJoCo simulator [42] is asked to grip spheres above a table. There are a total of 4 spheres and the robot receives a positive reward of $+1$ when its gripper catches a sphere (the sphere will disappear after being caught) otherwise it receives a time penalty. The maximal episode return is $+4$, and the minimal episode return is -1.

SuperMarioBros. A Mario agent with raw image inputs explores to discover the flag. The reward is based on the score given by the NES simulator [19] and is clipped into -1 and $+1$ except $+50$ for a flag. There are 24 stages in the SuperMarioBros game, but we only focus on the stages of 1–1, 1–2, and 1–3.

5.1   Comparison of Exploration Efficiency

In this section, we study the exploration efficiency in terms of the state distribution entropy. We focus on the Empty Room environment because it is tractable to calculate the state distribution entropy for this environment.

(a) Empty Room   (b) Four Rooms   (c) FetchReach   (d) SuperMarioBros

Fig. 2. Four environments considered in this paper. Four environments considered in this paper.

Table 1. Average entropy of visited state distribution over 5 random seeds on the Empty Room environment. Here we use ± to denote the standard deviation.

	Entropy
Random	5.129 ± 0.021
Bonus	5.138 ± 0.085
Novelty-pursuit	**5.285 ± 0.073**
Novelty-pursuit-planning-oracle	5.513 ± 0.077
Novelty-pursuit-counts-oracle	5.409 ± 0.059
Novelty-pursuit-oracles	5.627 ± 0.001
Maximum	5.666

We consider the following baselines: 1) random: uniformly selecting actions; 2) bonus: a curiosity-driven exploration method that uses the exploration bonus [7]; 3) novelty-pursuit: the proposed method. We also consider three variants of our method: 4) novelty-pursuit-planning-oracle: the proposed method with a perfect goal-conditioned planning policy; 5) novelty-pursuit-counts-oracle: the proposed method with goal-selection based on true visitation counts; 6) novelty-pursuit-oracles: the proposed method with the above two oracles. The results are summarized in Table 1. Here we measure the entropy over all visited states by the learned policy. Note that the maximum state distribution entropy for this environment is 5.666.

Firstly, we can see that novelty-pursuit achieves a higher entropy than the random and bonus method. Though the bonus method outperforms the random method, it is inefficient to a maximum state entropy exploration. We attribute this to delayed and imperfect feedbacks of the exploration bonus. Secondly, when the planning oracle and visitation counts oracle are available, the entropy of our method roughly improves by 0.228 and 0.124, respectively. We notice that the planning-oracle variant avoids random actions within the exploration boundary and spends more meaningful steps to explore around the exploration boundary, thus greatly improves the entropy. Based on this observation, we think accelerating goal-conditioned policy training is more important for our method. Thirdly, the combination of two oracles gives a near-perfect performance (the gap between the maximum state entropy is only 0.039). This result demonstrates that goal-condition exploration behaviors presented by novelty-pursuit can increase the state entropy.

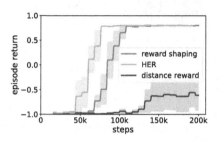

Fig. 3. Training curves of learned policies with different goal-conditioned policy training techniques on the Empty Room environment. Solid lines correspond to the mean of episode returns while shadow regions indicate the standard deviation.

5.2   Ablation Study of Training Techniques

In this section, we study the core component of our method regarding quickly learning a goal-conditioned policy.

In particular, we study the effects of qualities of goal-conditioned policies when using different training techniques. We compare HER and the reward-shaping with the distance reward function. Results on the Empty Room are shown in Fig. 3.

From Fig. 3, we find that both HER and reward shaping can accelerate training goal-conditioned policies. Similar to the planning-oracle in the previous section, such-trained goal-conditioned policy avoids random exploration within the explored region, hence it can quickly find the exit. The distance reward function may change the optimal behaviors of goal-conditioned policy and therefore does not perform well.

5.3   Evaluation on Complicated Environments

In this section, we compare different methods in terms of environment rewards. We will see that without sufficient and efficient exploration, the learned policy may be stuck into the local optimum. Two baseline methods are considered: 1) vanilla: DDPG [23] with Gaussian action noise on Fetch Reach with the continuous action space and ACER [43] with policy entropy regularization on other environments with the discrete action space; 2) bonus: a modified vanilla method that combines the environments reward and the exploration bonus [7]. Note reported results of novelty-pursuit are based on the performance of the exploitation policy π_e rather than the goal-conditioned exploration policy π_g. And we keep the same number of samples and policy optimization iterations for all methods to ensure fairness.

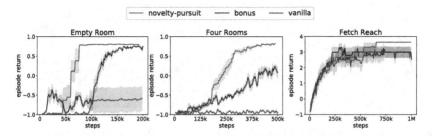

Fig. 4. Training curves of learned policies over 5 random seeds on the Empty Room, Four Rooms, and FetchReach environments. Solid lines correspond to the mean of episode returns while shadow regions indicate the standard deviation.

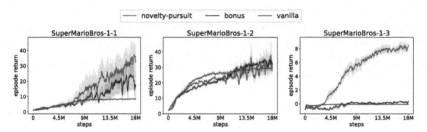

Fig. 5. Training curves of learned policies over 3 random seeds on the game of Super-MarioBros. Solid lines correspond to the mean of episode returns while shadow regions indicate the standard deviation.

Firstly, we consider the Empty Room and the Four Rooms environments. The results are shown in the first two parts of Fig. 4. We see that the vanilla method hardly finds the exit on the maze environments. Novelty-pursuit outperforms the bonus method on both environments. And we also observe that the behaviors of the bonus method are somewhat misled by the imperfect exploration bonus though we have tried many weights to balance the environment reward and exploration bonus.

Secondly, we consider the FetchReach environment, and results are shown in the most right part of Fig. 4. We see that novelty-pursuit can consistently grip 4 spheres while other methods sometimes fail to efficiently explore the whole state space to grip 4 spheres.

Finally, we consider the SuperMarioBros environments, in which it is very hard to discover the flag due to the huge state space and the long horizon. Learning curves are plotted in Fig. 5 and the final performance is listed in Table 2. We find the vanilla method gets stuck into the local optimum on SuperMarioBros-1-1 while the bonus method and ours can find a near-optimal policy. All methods perform well on SuperMarioBros-1-2 because of the dense rewards of this task. On SuperMarioBros-1-3, this task is very challenging because rewards are very sparse and all methods fail to discover the flag on this environment. To better understand the learned policies, we plot trajectories of different methods on

Fig. 6. Trajectory visualization on SuperMarioBros-1-3. Trajectories are plotted in green cycles with the same training samples (18 million). The agent starts from the most left part and needs to fetch the flag on the most right part. Top row: vanilla ACER; middle row: ACER + exploration bonus; bottom row: novelty-pursuit (ours).

SuperMarioBros-1-3 in Fig. 6, and more results on other environments can be found in Appendix A.2. It turns out only our method can get positive rewards and make certain progress via deep exploration behaviors presented by the goal-conditioned policy on SuperMarioBros-1-3.

Table 2. Final Performance of learned policies over 3 random seeds on SuperMario-Bros. We use ± to denote the standard deviation.

	Novelty-pursuit	Bonus	Vanilla
SuperMarioBros-1-1	**36.02 ± 8.19**	17.74 ± 7.84	8.43 ± 0.14
SuperMarioBros-1-2	**33.30 ± 6.13**	**33.19 ± 1.53**	29.64 ± 2.02
SuperMarioBros-1-3	**8.14 ± 0.55**	0.20 ± 0.14	−0.07 ± 0.01

6   Conclusion

We focus on the efficient exploration aspect of RL in this paper. We propose a goal-section criterion based on the principle of maximum state entropy exploration and demonstrate the proposed method is efficient towards exploring the whole state space. Therefore, the proposed method could escape from the local optimum and heads the (near-) optimal policy. Goal representation is somewhat manually extracted in our method and we believe an automatic representation learning for goal-conditioned learning can lead to more general applications. In addition to the way of off-policy learning, methods based on imitation learning to leverage good experience is another promising direction.

Acknowledgements. The authors will thank Fabio Pardo for sharing ideas to visualize trajectories for SuperMarioBros. In addition, the authors appreciate the helpful instruction from Dr. Yang Yu and the insightful discussion with Tian Xu as well as Xianghan Kong.

A   Appendix

A.1   Reward Shaping for Training Goal-Conditioned Policy

Reward shaping is invariant to the optimal policy under some conditions [27]. Here we verify that the reward shaping function introduced by our method doesn't change the optimal behaviors for goal-conditioned policy. Lets' consider the total shaping rewards during an episode of length T:

$$\sum_{t=1}^{T} -d(ag_t, g) + d(ag_{t+1}, g)$$
$$= -d(ag_1, g) + d(ag_2, g) - d(ag_2, g) + d(ag_3, g) \cdots$$
$$= -d(ag_1, g) + d(ag_{T+1}, g)$$

For the optimal policy π_g^*, $d(ag_{T+1}, g) = 0$ while $d(ag_1, g)$ is a constant. Therefore, the optimal policy π_g induced by the reward shaping is invariant to the one induced by the sparse reward function in Eq. 2.

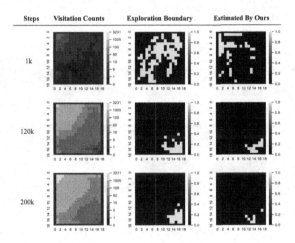

Fig. 7. Visualization for the true visitation counts and the corresponding exploration boundary.

A.2   Additional Results

In this part, we provide additional experiment results to better understand our method.

Empty Room. We visualize the true visitation counts and the corresponding exploration boundary in Fig. 7. Note the agent starts from the left top corner and the exit is on the most bottom right corner. The data used for visualization is collect by a random policy. Hence, the visitation counts are large on the left top part. We define the true exploration boundary as the top 10% states with least visitation counts and the estimated exploration boundary given by our method are states with the largest prediction errors in the priority queue. From this figure, we can see that our method can make a good approximation to the true exploration boundary given by visitation counts.

SuperMarioBros. In Fig. 8, we make additional trajectory visualization on SuperMarioBros-1-1 and SuperMarioBros-1-2. Trajectories are plotted with the same number of samples (18M). We can observe that the vanilla method gets into the local optimum on SuperMarioBros-1-1 even though it has used the policy entropy regularization to encourage exploration. In addition, only our method can get the flag on SuperMarioBros-1-2.

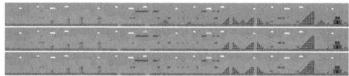

(a) SuperMarioBros-1-1. The agent starts from the most left part and needs to find the flag on the most right part.

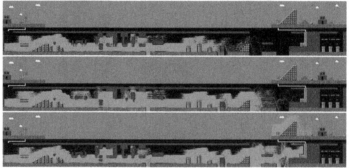

(b) SuperMarioBros-1-2. The agent starts from the most left part and needs to get the flag through the water pipe on the right part (see arrows).

Fig. 8. Trajectory visualization on SuperMarioBros-1-1 and SuperMarioBros-1-2. For each figure, top row: vanilla ACER; middle row: ACER + exploration bonus; bottom row: novelty-pursuit (ours). The vanilla method gets stuck into the local optimum on SuperMarioBros-1-1. Only our method can get the flag on SuperMarioBros-1-2.

A.3   Environment Prepossessing

In this part, we present the used environment preprocessing.

Maze. Different from [9], we only use the image and coordination information as inputs. Also, we only consider four actions: turn left, turn right, move forward and move backward. The maximal episode length is 190 for Empty Room, and 500 for Four Rooms. Each time the agent receives a time penalty of 1/max_episode_length and receives a reward of +1 when it finds the exit.

FetchReach. We implement this environment based on *FetchReach-v0* in Gym [5]. The maximal episode length is 50. The xyz coordinates of four spheres are $(1.20, 0.90, 0.65)$, $(1.10, 0.72, 0.45)$, $(1.20, 0.50, 0.60)$, and $(1.45, 0.50, 0.55)$. When sampling goals, we resample goals if the target position is outside of the table i.e., the valid x range: $(1.0, 1.5)$, the valid y range is $(0.45, 1.05)$, and the valid z range is $(0.45, 0.65)$.

SuperMarioBros. We implement this environment based on [19] with OpenAI Gym wrappers. Prepossessing includes grey-scaling, observation downsampling, external reward clipping (except that 50 for getting flag), stacked frames of 4, and sticky actions with a probability of 0.25 [26]. The maximal episode length is 800. The environment restarts to the origin when the agent dies.

A.4 Network Architecture

We use the convolutional neural network (CNN) for Empty Room, Four Rooms, and video games of SuperMarioBros, and multi-layer perceptron (MLP) for FetchReach environment. Network architecture design and parameters are based on the default implementation in OpenAI baselines [11]. For each environment, RND uses a similar network architecture. However, the predictor network has additional MLP layers than the predictor network to strengthen its representation power [7].

A.5 Hyperparameters

Table 3 gives hyperparameters for ACER [43] on the maze and SuperMarioBros (the learning algorithm is RMSProp [41]. DDPG [23] used in Fetch Reach environments is based on the HER algorithm implemented in OpenAI baselines [11] expect that the actor learning rate is 0.0005. We run 4 parallel environments for DDPG and the size of the priority queue is also 100. As for the predictor network, the learning rate of the predictor network is 0.0005 and the optimization algorithm is Adam [20] for all experiments, and the batch size of training data is equal to the product of rollout length and the number of parallel environments.

The goal-conditioned exploration policy of our method is trained by combing the shaping rewards defined in Eq. 3 and environment rewards, which helps reduce the discrepancy with the exploitation policy. The weight for environment rewards is 1 for all environments except 2 for SuperMarioBros. For the bonus method used in Sect. 5, the weight β to balance the exploration bonus is 0.1 for Empty Room and Four Rooms, 0.01 for FetchReach, 1.0 for SuperMarioBros-1-1 and SuperMarioBros-1-3, and 0.1 for SuperMarioBros-1-2. Following [6,7] we also do a normalization for the exploration bonus by dividing them via a running estimate of the standard deviation of the sum of discounted exploration bonus. In addition, we find sometimes applying the imitation learning technique for the goal-conditioned policy can improve performance. We will examine this in detail in future works. Though we empirically find HER is useful in simple environments like the maze and the MuJoCo robotics tasks, we find it is less powerful than the technique of reward shaping on complicated tasks like SuperMarioBros. Hence, reported episode returns of learned policies are based on the technique of reward shaping on SuperMarioBros and HER for others.

For the exploitation policy, we periodically allow it to interact with the environment to mitigate the exploration error [16]. For all experiments, we split the half interactions for the exploitation method. For example, if the number of maximal samples is $200k$, the exploration and the exploitation policy will use the same $100k$ interactions.

Table 3. Hyperparameters of our method based on ACER on the maze and Super-MarioBros environments.

Hyperparameters	Empty room	Four rooms	SuperMarioBros
Rollout length	20	20	20
Number of parallel environments	4	4	8
Learning rate	0.0007	0.0007	0.00025
Learning rate schedule	Linear	Linear	Constant
Discount factor γ	0.95	0.95	0.95
Entropy coefficient	0.10	0.10	0.10
Size of priority queue	100	100	20
Total training steps	200K	500K	18M

References

1. Andrychowicz, M., et al.: Hindsight experience replay. In: Proceedings of the 30th Annual Conference on Neural Information Processing Systems, pp. 5048–5058 (2017)
2. Auer, P., Cesa-Bianchi, N., Fischer, P.: Finite-time analysis of the multiarmed bandit problem. Machine Learning **47**(2–3), 235–256 (2002). https://doi.org/10.1023/A:1013689704352
3. Baranes, A., Oudeyer, P.: R-IAC: robust intrinsically motivated exploration and active learning. IEEE Trans. Auton. Ment. Dev. **1**(3), 155–169 (2009)
4. Bellemare, M.G., Srinivasan, S., Ostrovski, G., Schaul, T., Saxton, D., Munos, R.: Unifying count-based exploration and intrinsic motivation. In: Proceedings of the 29th Annual Conference on Neural Information Processing Systems, pp. 1471–1479 (2016)
5. Brockman, G., et al.: OpenAI gym. CoRR **1606**, 01540 (2016)
6. Burda, Y., Edwards, H., Pathak, D., Storkey, A.J., Darrell, T., Efros, A.A.: Large-scale study of curiosity-driven learning. In: Proceedings of the 7th International Conference on Learning Representations (2019)

7. Burda, Y., Edwards, H., Storkey, A.J., Klimov, O.: Exploration by random network distillation. In: Proceedings of the 7th International Conference on Learning Representations (2019)
8. Chen, X., Yu, Y.: Reinforcement learning with derivative-free exploration. In: Proceedings of the 18th International Conference on Autonomous Agents and Multi-Agent Systems, pp. 1880–1882 (2019)
9. Chevalier-Boisvert, M., Willems, L., Pal, S.: Minimalistic gridworld environment for openai gym. https://github.com/maximecb/gym-minigrid (2018)
10. Choromanski, K., Pacchiano, A., Parker-Holder, J., Tang, Y., Sindhwani, V.: From complexity to simplicity: Adaptive es-active subspaces for blackbox optimization. In: Proceedings of the 32nd Annual Conference on Neural Information Processing Systems, pp. 10299–10309 (2019)
11. Dhariwal, P., et al.: OpenAI baselines (2017). https://github.com/openai/baselines
12. Ecoffet, A., Huizinga, J., Lehman, J., Stanley, K.O., Clune, J.: Go-explore: a new approach for hard-exploration problems. CoRR 1901.10995 (2019)
13. Florensa, C., Held, D., Geng, X., Abbeel, P.: Automatic goal generation for reinforcement learning agents. In: Proceedings of the 35th International Conference on Machine Learning, pp. 1514–1523 (2018)
14. Forestier, S., Mollard, Y., Oudeyer, P.: Intrinsically motivated goal exploration processes with automatic curriculum learning. CoRR 1708.02190 (2017)
15. Fortunato, M., et al.: Noisy networks for exploration. In: Proceedings of the 6th International Conference on Learning Representations (2018)
16. Fujimoto, S., Meger, D., Precup, D.: Off-policy deep reinforcement learning without exploration. In: Proceedings of the 36th International Conference on Machine Learning, pp. 2052–2062 (2019)
17. Haarnoja, T., Zhou, A., Abbeel, P., Levine, S.: Soft actor-critic: Off-policy maximum entropy deep reinforcement learning with a stochastic actor. In: Proceedings of the 35th International Conference on Machine Learning, pp. 1856–1865 (2018)
18. Hazan, E., Kakade, S.M., Singh, K., Soest, A.V.: Provably efficient maximum entropy exploration. In: Proceedings of the 36th International Conference on Machine Learning, pp. 2681–2691 (2019)
19. Kauten, C.: Super Mario Bros for OpenAI Gym (2018). https://github.com/Kautenja/gym-super-mario-bros
20. Kingma, D.P., Ba, J.: Adam: A method for stochastic optimization. In: Proceedings of the 3rd International Conference on Learning Representations (2015)
21. Kolter, J.Z., Ng, A.Y.: Near-bayesian exploration in polynomial time. In: Proceedings of the 26th International Conference on Machine Learning, pp. 513–520 (2009)
22. Lattimore, T., Hutter, M.: Near-optimal PAC bounds for discounted MDPs. Theor. Comput. Sci. **558**, 125–143 (2014)
23. Lillicrap, T.P., et al.: Continuous control with deep reinforcement learning. In: Proceedings of the 4th International Conference on Learning Representations (2016)
24. Liu, F., Li, Z., Qian, C.: Self-guided evolution strategies with historical estimated gradients. In: Proceedings of the 29th International Joint Conference on Artificial Intelligence, pp. 1474–1480 (2020)
25. Mnih, V., et al.: Asynchronous methods for deep reinforcement learning. In: Proceedings of the 33rd International Conference on Machine Learning, pp. 1928–1937 (2016)
26. Mnih, V., et al.: Human-level control through deep reinforcement learning. Nature **518**(7540), 529–533 (2015)

27. Ng, A.Y., Harada, D., Russell, S.J.: Policy invariance under reward transformations: Theory and application to reward shaping. In: Proceedings of the 16th International Conference on Machine Learning (1999)
28. O'Donoghue, B., Munos, R., Kavukcuoglu, K., Mnih, V.: PGQ: combining policy gradient and q-learning. CoRR 1611.01626 (2016)
29. Osband, I., Blundell, C., Pritzel, A., Roy, B.V.: Deep exploration via bootstrapped DQN. In: Proceedings of the 29th Annual Conference on Neural Information Processing Systems, pp. 4026–4034 (2016)
30. Ostrovski, G., Bellemare, M.G., van den Oord, A., Munos, R.: Count-based exploration with neural density models. In: Proceedings of the 34th International Conference on Machine Learning, pp. 2721–2730 (2017)
31. Pathak, D., Agrawal, P., Efros, A.A., Darrell, T.: Curiosity-driven exploration by self-supervised prediction. In: Proceedings of the 34th International Conference on Machine Learning, pp. 2778–2787 (2017)
32. Péré, A., Forestier, S., Sigaud, O., Oudeyer, P.: Unsupervised learning of goal spaces for intrinsically motivated goal exploration. In: Proceedings of the 6th International Conference on Learning Representations (2018)
33. Plappert, M., et al.: Multi-goal reinforcement learning: Challenging robotics environments and request for research. CoRR 1802.09464 (2018)
34. Plappert, M., et al.: Parameter space noise for exploration. In: Proceedings of the 6th International Conference on Learning Representations (2018)
35. Schaul, T., Horgan, D., Gregor, K., Silver, D.: Universal value function approximators. In: Proceedings of the 32nd International Conference on Machine Learning, pp. 1312–1320 (2015)
36. Schulman, J., Wolski, F., Dhariwal, P., Radford, A., Klimov, O.: Proximal policy optimization algorithms. CoRR **1707**, 06347 (2017)
37. Stadie, B.C., Levine, S., Abbeel, P.: Incentivizing exploration in reinforcement learning with deep predictive models. CoRR 1507.00814 (2015)
38. Strehl, A.L., Littman, M.L.: An analysis of model-based interval estimation for markov decision processes. J. Comput. Syst. Sci. **74**(8), 1309–1331 (2008)
39. Sutton, R.S., Barto, A.G.: Introduction to Reinforcement Learning. MIT Press, Cambridge (1998)
40. Szepesvári, C.: Algorithms for Reinforcement Learning. Morgan & Claypool Publishers, New York (2010)
41. Tieleman, T., Hinton, G.: Lecture 6.5-rmsprop: Divide the gradient by a running average of its recent magnitude. COURSERA: Neural Netw. Mach. Learn. **4**(2), 26-31 (2012)
42. Todorov, E., Erez, T., Tassa, Y.: MuJoCo: a physics engine for model-based control. In: 2012 IEEE/RSJ International Conference on Intelligent Robots and Systems, pp. 5026–5033 (2012)
43. Wang, Z., et al.: Sample efficient actor-critic with experience replay. In: Proceedings of the 5th International Conference on Learning Representations (2017)
44. Yu, Y.: Towards sample efficient reinforcement learning. In: Proceedings of the 27th International Joint Conference on Artificial Intelligence, pp. 5739–5743 (2018)

Context-Aware Multi-agent Coordination with Loose Couplings and Repeated Interaction

Feifei Lin$^{(\boxtimes)}$, Xu He, and Bo An

Nanyang Technological University, Singapore, Singapore
{linf0012,hexu0003,boan}@ntu.edu.sg

Abstract. Coordination between multiple agents can be found in many areas of industry or society. Despite a few recent advances, this problem remains challenging due to its combinatorial nature. First, with an exponentially scaling action set, it is challenging to search effectively and find the right balance between exploration and exploitation. Second, performing maximization over all agents' actions jointly is computationally intractable. To tackle these challenges, we exploit the side information and loose couplings, i.e., conditional independence between agents, which is often available in coordination tasks. We make several key contributions in this paper. First, the repeated multi-agent coordination problem is formulated as a multi-agent contextual bandit problem to balance the exploration-exploitation trade-off. Second, a novel algorithm called MACUCB is proposed, which uses a modified zooming technique to improve the context exploitation process and a variable elimination technique to efficiently perform the maximization through exploiting the loose couplings. Third, two enhancements to MACUCB are proposed with improved theoretical guarantees. Fourth, we derive theoretical bounds on the regrets of each of the algorithms. Finally, to demonstrate the effectiveness of our methods, we apply MACUCB and its variants to a realistic cloudlet resource rental problem. In this problem, cloudlets must coordinate their computation resources in order to optimize the quality of service at a low cost. We evaluate our approaches on a real-world dataset and the results show that MACUCB and its variants significantly outperform other benchmarks.

Keywords: Multi-agent contextual bandit · Multi-agent coordination · Loose couplings · Cloudlet computing

1 Introduction

Many real-life problems could be considered as multi-agent coordination problems, which require agents to coordinate their actions repeatedly to optimize a

Supported by the Alibaba-NTU Singapore Joint Research Institute, Nanyang Technological University.

global utility. It is an important issue in multi-agent systems, with a wide range of application domains. Examples include robotic systems [12], traffic light control [22] and maintenance planning [19]. However, the size of the joint action set scales exponentially with the number of agents. Thus, how to optimally coordinate in repeated settings becomes an extremely challenging task for several reasons: First, as agents are unaware of the expected payoffs associated with different joint actions, they must explore to learn these values. With a large joint action set, inefficient algorithms will spend a large proportion of time in exploring sub-optimal actions, resulting in high regret. Second, it is not trivial to perform the optimization over an exponentially growing action set since computation and storage grow exponentially in the number of agents.

Fortunately, loose couplings exist in many coordination problems, meaning that each agent's action only has a direct impact on a subset of adjacent agents. Therefore, the global utility can break down into local utilities that only depend upon a small subset of agents. In addition, most real-life applications have side information, which can be highly informative of which type of actions should be taken in the future, especially when the action set is very large. Thus, we are interested in multi-agent coordination problems where a set of loosely coupled agents repeatedly observe state (side information) and have to perform actions jointly such that the expected global utility is maximized.

Multi-agent coordination problems have long been of great interest given their importance. Some reinforcement learning studies also consider coordination problems with loose couplings [8,10,13,18]. However, reinforcement learning focuses on sequential decision-making problems while ours is a single-stage setting. Moreover, most model-free learning works only concentrate on empirical results with no theoretical guarantee [13]. Our work is most relevant to [3,21], which also exploit the loose couplings in multi-agent multi-armed settings. However, their works fail to link side information with rewards of actions, neither do they exploit the similarities across agents. Our learning problem is of a combinatorial nature. In this sense, it is related to combinatorial bandits, which extend the classical multi-armed bandit (MAB) framework by allowing multiple plays at each round [4,5,7,9]. Similar to our settings, Qin et al. [15] study a contextual combinatorial bandits, with semi-bandit feedback [1], where action space grows exponentially and side information as well as the outcomes of all actions played are observable [15]. However, their work does not restrict the set of actions played, making it inapplicable to our problem where each agent can only play a single action from its own action set. Different from previous works, we exploit the side information and loose couplings to address these issues and provide several key contributions. First, we formulate the repeated multi-agent coordination problem as a multi-agent contextual bandit problem to balance the exploration-exploitation tradeoff in the joint actions of multiple agents. Second, we present a novel algorithm called MACUCB, which combines a modified zooming technique [20] and a variable elimination algorithm [3,16,17] to adaptively exploit the context information and address the unavoidable scalability issues in multi-agent settings. Third, we propose two enhancements to our base algorithm with improved bound guarantees. One is to share context space among agents and the other algorithm takes advantage of the full feedback information. Fourth, we show that the regret of MACUCB and its two variants are bounded. Finally, we empirically compare our algorithms with other state-of-the-art methods in a cloudlet resource rental problem and show that MACUCB and its variants achieve much lower empirical regret.

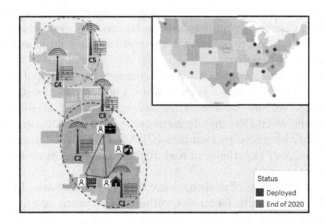

Fig. 1. A map of the United States showing cities with cloudlet infrastructures (upper-right corner). A map of Chicago with cloudlet Deployment locations (left)

2 Motivation Scenario

In this section, we use the cloudlet resource rental problem as a motivating example, while our model can be applied to a variety of multi-agent coordination scenarios.

Although mobile devices are getting more powerful recently, they still fall short to execute complex rich media applications like Pokémon Go. Computing offloading through the cloud is an effective way to solve this problem. However, cloud servers are usually located in the far distance, resulting in high latencies. In such context, cloudlets, deployed geographically near mobile users, have been proposed to provide services with low-latency responses. Foreseeing tremendous opportunities, many companies are expanding their investments in this field. For example, as shown in the upper-right corner of Fig. 1, Vapor IO will have its Kinetic Edge live in 20 US metropolitan markets. Now assume Niantic, the application service provider (ASP) of Pokémon Go has decided to rent computation resources to deploy its application in Chicago. Let us see how the user experience of Pokémon Go players in Chicago will be affected by the rental decisions. As depicted in Fig. 2, many players require for cloudlet services at the same time. Then, each cloudlet needs to decide how much resources to rent considering the side information e.g. past user demand pattern and the current time.

For example, assume that ASP makes rental decisions as shown in Fig. 2. Since the ASP rents sufficient computation resources, the computation tasks of users 1 to 3 are offloaded to the Cloudlet 1, leading to low latency. Therefore, the user experience at Cloudlet 1 is high. However, due to limited or no computing resources rented at Cloudlet 2 and Cloudlet n, the tasks of mobile users 5, $k-2$ to k are rejected. Then these tasks have to be offloaded to Cloud via a macro base station (MBS) through congested backbone Internet (dash line), resulting in high service delay and bad user experience at these cloudlets. Therefore, joint rental decisions at multiple cloudlets have to be carefully decided in order to get excellent overall user experience with a low cost, which is measured in terms of time-saving in task processing.

Note that cloudlets are in fact loosely coupled, thus the global utility can be decomposed into local ones. For example, Fig. 1 shows the daily activities of mobile user 1. As can be seen, this user spends most of the time in the blue region. Thus, given that mobile users can only access the nearest cloudlet, he/she will mostly be served by cloudlets (C1, C2, C3) located in that particular region. Therefore, user experience in each region only depends on a small subset of cloudlets.

However, there are lots of challenges involved in this scenario: First, the user demand, which is the main factor determining the benefit of rental decisions, is unknown ahead of time. Second, service demand and resource rental options might be different at different cloudlets as the market size and demand elasticity often vary across geographic locations. Thus, simply treating these cloudlets as a single agent might not work well. Third, the number of joint rental options increase exponentially with the number of cloudlets, making it not trivial to compute the optimal solution.

Fortunately, side information that combines contextual knowledge with historical data could be highly informative in the prediction of the future. Thus, in this paper, we model the scenario as a multi-agent contextual bandit problem and take advantage of the side information and loose couplings to address these challenges.

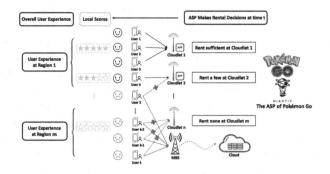

Fig. 2. An example of the cloudlet resource rental problem

3   Problem Description

We consider repeated interactions for a horizon of T rounds and the computation resource rental problem can be modelled as a tuple $\mathcal{N} = \langle \mathcal{C}, \mathcal{X}, \mathcal{A}, \mathcal{S}, \mathcal{F} \rangle$

- $\mathcal{C} = \{i\}$ ($|\mathcal{C}| = n$) is the set of n agents (e.g. cloudlets).
- $\mathcal{X} = \mathcal{X}_1 \times \cdots \times \mathcal{X}_n$ is the joint context space, which is the cross-product of the context space of individual context space $\mathcal{X}_i = \{x_i\}$. The joint context at time t is denoted by $\mathbf{x}_t = (x_{1,t}, \ldots, x_{n,t})$. In the cloudlet resource rental problem, side information $x_{i,t}$ can be user factor (e.g. past demand patterns), temporal factor (e.g.. current time) or other relevant factors related to cloudlet i in round t.
- $\mathcal{A} = \mathcal{A}_1 \times \cdots \times \mathcal{A}_n$ is the joint action set, defined as the cross-product of the individual action sets $\mathcal{A}_i = \{a_i\}$. The joint action at time t is denoted by $\mathbf{a}_t = (a_{1,t}, \ldots, a_{n,t})$. In the resource rental problem, an action $a_i \in \mathcal{A}_i$ denotes the number of virtual machines rented at cloudlet i.
- $\mathcal{S}$ is the individual score function. For each agent i, $\mathcal{S}_i$ is defined on context and action set, i.e., $\mathcal{S}_i : \mathcal{A}_i \times \mathcal{X}_i \rightarrow [0, 1]$. The observed value of the score is denoted by $s(a_{i,t}, x_{i,t})$ and its expected value by $\mu(a_{i,t}, x_{i,t}) = \mathbb{E}\left(s(a_{i,t}, x_{i,t})\right)$. In the motivation scenario, it evaluates the service quality of cloudlet i (in terms of achieved delay reduction) minus rental cost.
- $\mathcal{F}$ measures the expected global utility. In particular, in the motivation scenario, it measures the improvement in service quality minus the cost associated with the rental decisions.

In this paper, we consider multi-agent coordination problems in which the expected global utility function satisfies two properties. Firstly, the expected global utility can be represented as a function of the joint action and the score expectation vector, i.e., $\mathcal{F}(\mathbf{a}, \boldsymbol{\mu})$, where $\boldsymbol{\mu} = \left\{\{\mu(a_i, x_i)\}_{a_i \in \mathcal{A}_i}\right\}_{i \in \mathcal{C}}$ denotes the score expectation vector of all actions of all agents in set $\mathcal{C}$. More specifically, the expected utility at time t is $\mathcal{F}(\mathbf{a}_t, \boldsymbol{\mu}_t) = \mathcal{F}\left(\{\mu(a_{i,t}, x_{i,t})\}_{i \in \mathcal{C}}\right)$, where $\{\mu(a_{i,t}, x_{i,t})\}_{i \in \mathcal{C}}$ is an n-dimensional vector restricted on the action $a_{i,t}$ taken by each agent i. For example, as shown in Fig. 2, the expected overall user experience is determined by the rental decisions and the corresponding expected local scores obtained at each cloudlet.

Secondly, since agents are loosely coupled, they could be decomposed into m possible overlapping subsets $\mathcal{C}^j$. Correspondingly, the expected global utility could break down into m local expected utility functions f^j, i.e., $\mathcal{F}(\mathbf{a}, \boldsymbol{\mu}) = \sum_{j=1}^{m} f^j(\mathbf{a}^j, \boldsymbol{\mu}^j)$, where $\mathbf{a}^j$ and $\boldsymbol{\mu}^j$ are the local joint action and local score expectation vector respectively. For instance, as mentioned in the motivation scenario, the overall experience can be decomposed into regional ones. Often, the loose couplings structure can be illustrated using a coordination graph (CoG) [10,13].

In each round t, agents observe the joint context $\mathbf{x}_t$ and are asked to choose a joint action $\mathbf{a}_t$. Once the decision is made, agents observe the local scores $\{s(a_{i,t}, x_{i,t})\}_{i \in \mathcal{C}}$ and receive a global utility. The objective is to maximize the expected cumulative global utility $\sum_{t=1}^{T} \mathcal{F}(\mathbf{a}_t, \boldsymbol{\mu}_t)$ over T rounds. It is equivalent to minimize the expected cumulative regret Reg_T, defined as the difference in cumulative global utility between the joint actions we selected and the best actions $\mathbf{a}_t^*$, where $\mathbf{a}_t^* \triangleq \arg\max_{\mathbf{a} \in \mathcal{A}} \mathcal{F}(\mathbf{a}, \boldsymbol{\mu}_t)$. Let $\mathbf{a}_t^{j^*} = \{a_{i,t}^*\}_{i \in \mathcal{C}^j}$ be the best action restricted on agents in set C^j. Then the objective is to minimize the expected cumulative regret Reg_T

$$Reg_T = \sum_{t=1}^{T} \sum_{j=1}^{m} f^j(\mathbf{a}_t^{j^*}, \boldsymbol{\mu}_t^j) - f^j(\mathbf{a}_t^j, \boldsymbol{\mu}_t^j)$$

Before carrying out our analysis, let us make some natural assumptions about score functions and utility functions.

Lipschitz Score Functions. Consider a metric space of context of any agent i $(\mathcal{X}_i, \mathcal{D})$, where $\mathcal{D}$ defines the distance on $\mathcal{X}_i$. The local score function is Lipschitz with respect to metric $\mathcal{D}$. More specifically, $\forall i, \forall a_{i,t} \in \mathcal{A}_i$, and $\forall \mathbf{v}, \mathbf{w} \in \mathcal{X}_i$, score function $\mathcal{S}_i$ satisfies

$$\left| \mathcal{S}_i(a_{i,t}, \mathbf{v}) - \mathcal{S}_i(a_{i,t}, \mathbf{w}) \right| \leq \mathcal{D}(\mathbf{v}, \mathbf{w}) \tag{1}$$

Without loss of generality, we assume that the diameter of $\mathcal{X}_i$ is not more than 1, i.e., $\forall i, \sup_{\mathbf{u}, \mathbf{v} \in \mathcal{X}_i} \mathcal{D}(\mathbf{u}, \mathbf{v}) \leq 1$. Consider the motivation scenario. It is natural to assume that, when renting the same number of virtual machines, cloudlets with similar contexts (e.g.. demand patterns) have similar score ratings.

On the other hand, the expected utility functions depend on the actual problem instance. It might be simply the sum of the expected scores of actions $\mathbf{a}^j$ taken by agents in $\mathcal{C}^j$, i.e., $f^j(\mathbf{a}_t^j, \boldsymbol{\mu}_t^j) = \sum_{i \in \mathcal{C}^j} \mu(a_{i,t}, x_{i,t})$. It might also be complicated non-linear utilities. We simply assume that the expected reward f^j satisfies the following two assumptions.

Lipschitz Utility Functions. The expected local utility functions f^j is Lipschitz continuous with respect to the score expectation vector $\boldsymbol{\mu}^j$. In particular, there exists a universal constant $\alpha > 0$ such that, for $\forall j$ and any two score expectation vector $\hat{\boldsymbol{\mu}}^j$ and $\tilde{\boldsymbol{\mu}}^j$, we have

$$\left| f^j(\mathbf{a}_t^j, \hat{\boldsymbol{\mu}}_t^j) - f^j(\mathbf{a}_t^j, \tilde{\boldsymbol{\mu}}_t^j) \right| \leq \alpha \sum_{i \in \mathcal{C}^j} \left| \hat{\mu}(a_{i,t}, x_{i,t}) - \tilde{\mu}(a_{i,t}, x_{i,t}) \right| \tag{2}$$

Take the cloudlet rental problem as an example. It is intuitive that similar score ratings at cloudlets lead to similar regional user experience and vice versa.

Monotonic Utility Functions. The expected utility functions f^j is monotone non-decreasing with respect to the score expectation vector $\boldsymbol{\mu}^j$. Formally, for $\forall \mathbf{a}_t^j \in \mathcal{A}^j$, if $\hat{\mu}(a_{i,t}, x_{i,t}) \leq \tilde{\mu}(a_{i,t}, x_{i,t})$ for $\forall i \in \mathcal{C}^j$, we have

$$f^j(\mathbf{a}_t^j, \hat{\boldsymbol{\mu}}_t^j) \leq f^j(\mathbf{a}_t^j, \tilde{\boldsymbol{\mu}}_t^j) \tag{3}$$

The intuition behind the assumption is that the user experience definitely improves when score ratings at all cloudlets become higher. Additionally, it is not necessary for cloudlets to possess a direct knowledge of how the expected local utility functions $f^j\left(\mathbf{a}^j, \boldsymbol{\mu}^j\right)$ are defined. Instead, we assume there is an oracle, which takes joint action $\mathbf{a}$ and expected score $\boldsymbol{\mu}$ as input, and outputs the value of expected utilities $f^j\left(\mathbf{a}^j, \boldsymbol{\mu}^j\right)$.

4 Algorithms

Algorithm 1: MACUCB

---

1 **for** *each agent i* **do**
2   **for** *each action $a \in \mathcal{A}_i$* **do**
3         $B_{i,a} \leftarrow B(o, 1)$ where o is an arbitrary centre
4         $n(B_{i,a}) = 0$; $\mathcal{B}^1_{i,a} \leftarrow \{B_{i,a}\}$

5 **for** $t = 1, \ldots, T$ **do**
6   **for** *each agent i* **do**
7   Input context $x_{i,t}$
8         $Relevant \leftarrow \{B \in \mathcal{B}^t_i : x_{i,t} \in dom_t(B)\}$
9   **for** *each $B_{i,a} \in Relevant$* **do**
10   Calculate B_{rep} by Eq. (4) and update $s_t(B_{i,a})$ and $U_t(B_{i,a})$ by Eq. (5)
11   **for** *each action $a \in \mathcal{A}_i$* **do**
12   Calculate $\hat{B}_{i,a}$ by Eq. (6) and update $\hat{\mu}_t(a, x_{i,t})$ and $U_t(a)$ by Eq. (7)

13   Calculate a_t by Eq. (8)
14   Execute a_t and observe local scores $\{s(x_{i,t}, a_{i,t})\}_{i \in \mathcal{C}}$
15   **for** *each agent i* **do**
16         $n_{t+1}(\hat{B}_{i,a_{i,t}}) = n_t(\hat{B}_{i,a_{i,t}}) + 1$
17         $sum(\hat{B}_{i,a_{i,t}}) = sum(\hat{B}_{i,a_{i,t}}) + s(x_{i,t}, a_{i,t})$
18   **if** $conf_{t+1}(\hat{B}_{i,a_{i,t}}) \leq R(\hat{B}_{i,a_{i,t}})$ **then**
19             $B^{new} = B_{i,a_{i,t}}\left(x_{i,t}, \frac{1}{2}R(\hat{B}_{i,a_{i,t}})\right)$
20             $\mathcal{B}^{t+1}_{i,a_{i,t}} = \mathcal{B}^t_{i,a_{i,t}} \cup B^{new}$; $n_t(B^{new}) = 0$

4.1 Description of MACUCB

The basic idea of MACUCB is as follows: for each round t, the algorithm maintains a collection of balls $\mathcal{B}^t_i$ for each agent i, which forms a partition of the context space $\mathcal{X}_i$. Basically, each ball is a score estimator and the shape of balls guarantees that all context falling into the partition are within distance r from the centre. Thus, by Eq. (1), we can control the estimation errors by controlling the radius of the balls. Therefore, by generating more balls with smaller radius over time, our estimation becomes more accurate.

In detail, when context $x_{i,t}$ arrives, among all the balls whose domain contains $x_{i,t}$, the algorithm selects one ball $\hat{B}_{i,a}$ to estimate the score of each action

$a_i \in \mathcal{A}_i$ according to the **estimation rule**. Specifically, the estimation rule selects the ball (estimator) with the highest upper confidence bound. Then, based on the estimation, the algorithm plays the joint action $\mathbf{a}_t$ returned by the **selection rule** which also follows UCB criterion. Then the observed scores are used to update the estimation. In the end, a new ball with a smaller radius may be generated for each agent i according to the **generation rule** to give a refined partition when we are more confident about the estimation.

Now let us introduce some notations and definitions before stating the three rules. $\mathcal{B}_{i,a}^t$ denotes the collection of all balls associated with action a of agent i in round t. Moreover, define $\mathcal{B}_i^t$ as the set containing all balls of agent i in round t and $\mathcal{B}^t$ as the set containing balls of all agents in round t, i.e., $\mathcal{B}^t \triangleq \bigcup_{i \in \mathcal{C}} \mathcal{B}_i^t \triangleq \bigcup_{i \in \mathcal{C}} \left(\bigcup_{a \in \mathcal{A}_i} \mathcal{B}_{i,a}^t \right)$.

For action a of agent i, a ball with center o and radius r is defined by $B_{i,a}(o,r) = \{x \in \mathcal{X}_i : \mathcal{D}(x,o) \leq r\}$. For simplicity, it is abbreviated as $B_{i,a}$ in the subsequent sections. In addition, let $R(B_{i,a})$ denote the radius of ball $B_{i,a}$. Then the domain $\mathrm{dom}_t(B_{i,a})$ of the ball $B_{i,a}$ in round t is defined as a subset of $B_{i,a}$ that excludes all balls $B' \in \mathcal{B}_{i,a}^t$ with a smaller radius, i.e., $\mathrm{dom}_t(B_{i,a}) \triangleq B_{i,a} \backslash \left(\bigcup_{B'_{i,a} \in \mathcal{B}_{i,a}^t : R(B'_{i,a}) < R(B_{i,a})} B'_{i,a} \right)$.

Now we are ready to state the three rules.

Estimation Rule: The estimation rule has three steps.

(1) Basic Estimation. We say that $B_{i,a}$ is a **relevant** ball of agent i in round t if $x_{i,t} \in \mathrm{dom}_t(B_{i,a})$. For each ball $B_{i,a}$, it keeps two estimation statistics: the average score $\bar{s}_t(B_{i,a})$ and the confidence level $conf_t(B_{i,a})$. Let $n_t(B_{i,a})$ denote the number of rounds that $B_{i,a}$ has been selected before t and $sum(B_{i,a})$ be the sum of payoffs from these rounds. Then the average score $\bar{s}_t(B_{i,a})$ and the confidence level $conf_t(B_{i,a})$ are defined as

$$\bar{s}_t(B_{i,a}) \triangleq \frac{sum(B_{i,a})}{n_t(B_{i,a})}$$

$$conf_t(B_{i,a}) \triangleq \sqrt{\frac{4 \log T}{1 + n_t(B_{i,a})}}$$

(2) Refinement. To get a more accurate estimation, we perform a refinement for each relevant ball $B_{i,a}$, using statistics from the **representative ball** B_{rep} which gives an estimation with minimum uncertainty. It is defined as

$$B_{rep} \triangleq \arg\min_{B \in \mathcal{B}_{i,a}^t} D(B_{i,a}, B) + conf_t(B) + R(B) \tag{4}$$

Then the refinement is conducted as follows.

$$s_t(B_{i,a}) = \bar{s}_t(B_{rep})$$
$$U_t(B_{i,a}) = D(B_{rep}, B_{i,a}) + conf_t(B_{rep}) \tag{5}$$
$$+ R(B_{rep}) + R(B_{i,a})$$

where $s_t(B_{i,a})$ and $U_t(B_{i,a})$ represent the refined mean and confidence term respectively.

(3) UCB Estimation. After refinement, the ball $\hat{B}_{i,a}$ with the best upper confidence bound is selected to give the final estimation of the score. Specifically, for each action $a \in \mathcal{A}_i$, $\hat{B}_{i,a}$ is defined as

$$\hat{B}_{i,a} \triangleq \arg\max_{B_{i,a} \in Relevant} s_t(B_{i,a}) + U_t(B_{i,a}) \tag{6}$$

Then the final estimated mean score $\hat{\mu}_t(a, x_{i,t})$ and confidence term $U_t(a)$ are

$$\hat{\mu}_t(a, x_{i,t}) = s_t(\hat{B}_{i,a}) \text{ and } U_t(a) = U_t(\hat{B}_{i,a}) \tag{7}$$

The corresponding score expectation vector $\hat{\boldsymbol{\mu}}_t$ of all actions of all agents is $\hat{\boldsymbol{\mu}}_t = \left\{ \{\hat{\mu}_t(a, x_{i,t})\}_{a \in \mathcal{A}_i} \right\}_{i \in \mathcal{C}}$.

Selection Rule: The algorithm selects joint action and balls such that

$$\boldsymbol{a}_t = \arg\max_{\boldsymbol{a} \in \mathcal{A}} \mathcal{F}(\boldsymbol{a}, \hat{\boldsymbol{\mu}}_t) + C_t(\boldsymbol{a}) \tag{8}$$

where $\mathcal{F}(\boldsymbol{a}, \hat{\boldsymbol{\mu}}_t) = \sum_{j=1}^{m} f^j(\boldsymbol{a}^j, \hat{\boldsymbol{\mu}}_t^j)$ and $C_t(\boldsymbol{a}) = \alpha \sum_{j=1}^{m} \sum_{i \in \mathcal{C}^j} U_t(a_i)$. Then $\left\{ \hat{B}_{i,a_{i,t}} \right\}_{i \in \mathcal{C}}$ are the corresponding balls selected.

Generation Rule: For each agent i, if the ball selected $\hat{B}_{i,a_{i,t}}$ satisfies the inequality

$$conf_{t+1}(\hat{B}_{i,a_{i,t}}) \leq R(\hat{B}_{i,a_{i,t}}).$$

Then a new ball B^{new} is generated with center $x_{i,t}$ and radius $\frac{1}{2}R(\hat{B}_{i,a_{i,t}})$. We call $\hat{B}_{i,a_{i,t}}$ the **parent ball** of B^{new}.

The detail of the algorithm is presented in Algorithm 1. However, note that it is not trivial to calculate Eq. (8), since the joint action set $\mathcal{A}$ is exponential in the number of agents. Therefore, MACUCB calls a variable elimination algorithm (VE) [3, 16, 17] to perform this maximization.

4.2   Description of VE

The basic idea of VE is to exploit loose couplings to break down the problem into sub-problems, avoiding searching the whole joint action set. For each agent, we consider the sub-problem containing only local utility functions that have the agent in scope. Then agents are eliminated in sequence by calculating the value of agents' best responses to the neighbors in the sub-problems.

In details, let us rewrite the local utility functions (LUFs) f^j to include both the estimated utility and the confidence terms, i.e.,

$$f^j(\mathbf{a}^j) = f^j(\mathbf{a}^j, \hat{\boldsymbol{\mu}}^j) + \alpha \sum_{i \in \mathcal{C}^j} U(a_i)$$

By loose couplings, we can break down the problem into sub-problems. Let $\mathcal{F}$ be the set containing all LUFs, i.e., $\mathcal{F} \triangleq \{f^j\}_{j=1}^{m}$ and $\mathcal{F}_i$ be the set of LUFs that have agent i in scope. Then consider a subproblem that only has $\mathcal{F}_i$. Fixing a specific local joint action $\mathbf{a}^{\mathcal{C^j}-\mathbf{i}}$, the possible value $\mathcal{V}_i$ of $\mathcal{F}_i$ for all actions of agent i is

$$\mathcal{V}_i(\mathcal{F}_i, \mathbf{a}^{\mathcal{C^j}-\mathbf{i}}) = \bigcup_{a_i \in \mathcal{A}_i} \sum_{f^j \in \mathcal{F}_i} f^j(\mathbf{a}^{\mathcal{C^j}-\mathbf{i}} \times \{a_i\})$$

Since the action taken by agent i will only affect the global utility through $\mathcal{V}_i$, we can eliminate agent i by calculating the value of agent i 's best response to all joint action its neighbors can take $\mathbf{a}^{\mathcal{C^j}-\mathbf{i}} \in \mathcal{A}^{\mathcal{C^j}-i}$. Then a new LUF $\mathcal{F}_i^{new}$ can be constructed using these values, which depends only on $\mathbf{a}^{\mathcal{C^j}-\mathbf{i}}$, i.e.,

$$\mathcal{F}_i^{new}(\mathbf{a}^{\mathcal{C^j}-\mathbf{i}}) = \max\left(\mathcal{V}_i(\mathcal{F}_i, \mathbf{a}^{\mathcal{C^j}-\mathbf{i}})\right)$$

Then we replace $\mathcal{F}_i$ in $\mathcal{F}$ by the new factor. In addition, since we want to find the optimal joint action, we tag each value in $\mathcal{F}_i^{new}$ with the best response of agent i. We do these series of operations to eliminate all agents $i \in \mathcal{C}$ in a predetermined order q. Details of the algorithm are shown in Algorithm 2.

Algorithm 2: VE

---

1 **Input** A set of local utility functions f^j and an elimination order q containing all agents

2 $\mathcal{F} = \{f^j\}_{j=1}^m$

3 **while** q *is not empty* **do**

4 $\quad$ $i \leftarrow q.dequeue()$

5 $\quad$ **for** *each action* $\mathbf{a}^{\mathcal{C^j}-\mathbf{i}} \in \mathcal{A}^{\mathcal{C^j}-i}$ **do**

6 $\quad\quad$ $\mathcal{F}_i^{new}(\mathbf{a}^{\mathcal{C^j}-\mathbf{i}}) = \max\left(\mathcal{V}_i(\mathcal{F}_i, \mathbf{a}^{\mathcal{C^j}-\mathbf{i}})\right)$

7 $\quad$ **end**

8 $\quad$ $\mathcal{F} \leftarrow (\mathcal{F}\backslash\mathcal{F}_i) \cup \mathcal{F}_i^{new}$

9 **end**

10 $\mathbf{v} \leftarrow \mathcal{V}(\mathcal{F})$

11 **return** the tag a^* attached to $\mathbf{v}$

4.3   Extensions

The algorithm proposed for multi-agent contextual bandits can be naturally extended to the special case where agents have similar action sets, contextual spaces and score functions. In such cases, we treat $\mathcal{X} = \bigcup_{i \in \mathcal{C}} \mathcal{X}_i$ and $\mathcal{A} = \bigcup_{i \in \mathcal{C}} \mathcal{A}_i$ and modify MACUCB to allow information sharing between agents by sharing the same adaptive space partition $\mathcal{B}$. The new algorithm is named Multi-agent Similar Contextual Upper Confidence Bound (MASCUCB). For instance, in the cloudlet rental problem, cloudlets usually have similar rental options and score functions (since time delay is determined by demand and computation power regardless of locations). MASCUCB follows the general framework of MACUCB but instead of having an individual partition for each agent, it maintains a public collection of balls for all agents, i.e., $\mathcal{B}^t = \mathcal{B}_1^t = \cdots = \mathcal{B}_n^t$. Then, both the estimation and the update are done with the public collection, which greatly expedites the exploration process and leads to a decrease in regret. The update procedure is shown in Algorithm 3.

Algorithm 3: Update in MASCUCB

1 **for** *each agent i* **do**
2   Execute MACUCB Lines 16 - 17
3   **if** $conf(\hat{B}_{i,a_{i,t}}) \leq R(\hat{B}_{i,a_{i,t}})$ **then**
4          $B^{new} = B_{a_{i,t}}\left(x_{i,t}, \frac{1}{2}R(\hat{B}_{a_{i,t}})\right)$
5          $\mathcal{B}_{a_{i,t}}^{t+1} \leftarrow \mathcal{B}_{a_{i,t}}^t \cup B^{new}$; $n_t(B^{new}) = 0$

For some other applications, full feedback is available, i.e., scores of all actions are available at the end of each round. For example, in the cloudlet resource rental problem, the service demand in each region is revealed at the end of each round. Then the score, which is the delay reduction minus the rental cost, can be derived accordingly. Therefore, the scores of all rental decisions are observable. We call MASCUCB with full feedback as MASCUCBwF. As outlined in Algorithm 4, MASCUCBwF follows the general framework of MASCUCB, but updates balls for all actions $\{\hat{B}_{i,a}\}_{a_i \in \mathcal{A}_i}$ instead of only updating ball $\hat{B}_{i,a_{i,t}}$.

Algorithm 4: Update in MASCUCBwF

1 Execute a_t and observe local scores for all actions $\{s(a_i, x_{i,t})\}_{a_i \in \mathcal{A}_i, i \in \mathcal{C}}$
2 **for** *each agent i* **do**
3   **for** *each action a_i in $\mathcal{A}_i$* **do**
4          $n_{t+1}(\hat{B}_{i,a_i}) \leftarrow n_t(\hat{B}_{i,a_i}) + 1$
5          $sum(\hat{B}_{i,a_i}) = sum(\hat{B}_{i,a_i}) + s_t(x_{i,t}, a_i)$
6   **if** $conf(\hat{B}_{i,a_i}) \leq R(\hat{B}_{i,a_i})$ **then**
7              $B^{new} \leftarrow B_{a_i}\left(x_{i,t}, \frac{1}{2}R(\hat{B}_{i,a_i})\right)$
8              $\mathcal{B}_{a_i}^{t+1} \leftarrow \mathcal{B}_{a_i}^t \cup B^{new}$
9              $n_t(B^{new}) = 0$

---

5   Regret Analysis

In this section, we provide an upper bound on the cumulative regret for MACUCB.

Define the r-covering number of $(\mathcal{X}_i, D)$ as the minimal number of balls, whose diameters are not greater than r, needed to cover $\mathcal{X}_i$:

$$N_r(\mathcal{X}_i) \triangleq \min \left\{ H : \exists V = v_1, \ldots, v_H, \mathcal{X}_i \subset \bigcup_{h=1}^{H} B(v_h, \frac{r}{2}) \right\}$$

Then we have the following theorem regarding the expected regret achieved by MACUCB.

Theorem 1. *Assume Eqs. (1) and (2) hold. With probability at least $1 - 2nT^{-2}$, the expected global regret is bounded by*

$$Reg_T \leq 2\alpha \sum_{j=1}^{m} \sum_{i \in \mathcal{C}^j} \inf_{r_i' \in (0,1)} \left(7r_i'T + |\mathcal{A}_i| \sum_{r \in \mathbf{R}(r_i', 1)} \frac{28 N_r(\mathcal{X}_i) \log T}{r} \right)$$

where α is the Lipschitz continuity coefficient in Eq. (2) and $\mathbf{R}(a, b) = \{2^{-k} | k \in \mathbb{N} \wedge 2^{-k} \in (a, b]\}$.

Before we prove Theorem 1, let us first propose to bound the difference between the mean score $\mu(a, O_{B_{i,a}})$ of agent i's action a with context $O_{B_{i,a}}$ and the average score $\bar{s}_t(B_{i,a})$ for all balls. Define the **good event** be that

$$\forall t \in [T], \forall i \in \mathcal{C}, \forall a \in \mathcal{A}_i, \forall B_{i,a} \in \mathcal{B}_{i,a}^t,$$

$$\left| \bar{s}_t(B_{i,a}) - \mu(a, O_{B_{i,a}}) \right| \leq conf_t(B_{i,a}) + R(B_{i,a})$$

where $O_{B_{i,a}}$ denotes the centre of ball $B_{i,a}$ and $\mu(a, O_{B_{i,a}})$ is the mean score of action a with context $O_{B_{i,a}}$. The following lemma states that the good event happens with high probability.

Lemma 1. *Assume Eqs. (1) and (2) hold. For $\forall t \in [T], \forall i \in \mathcal{C}, \forall a \in \mathcal{A}_i, \forall B_{i,a} \in \mathcal{B}_{i,a}^t$, with probability at least $1 - 2nT^{-2}$, we have*

$$\left| \bar{s}_t(B_{i,a}) - \mu(a, O_{B_{i,a}}) \right| \leq conf_t(B_{i,a}) + R(B_{i,a})$$

Proof of Lemma 1. Fix a ball $B_{i,a}$. If $n_t(B_{i,a}) = 0$ or $R(B_{i,a}) = 1$, we have

$$\left| \bar{s}_t(B_{i,a}) - \mu(a, O_{B_{i,a}}) \right| \leq 1 \leq conf_t(B_{i,a}) + R(B_{i,a})$$

Thus, the inequality always holds if $n_t(B_{i,a}) = 0$ or $R(B_{i,a}) = 1$.
If $n_t(B_{i,a}) \geq 1$ and $R(B_{i,a}) < 1$, by Eq. (1) we have

$$\left| \mathbf{E} \left[\bar{s}_t(B_{i,a}) \right] - \mu \left(a, O_{B_{i,a}} \right) \right| \leq \max_{x_{i,t} \in dom_t(B_{i,a})} \mathcal{D}(x_{i,t}, O_{B_{i,a}}) \leq R(B_{i,a})$$

Therefore, according to Hoeffding's inequality,

$$\Pr\Big(\big| \bar{s}_t(B_{i,a}) - \mu(a, O_{B_{i,a}}) \big| > conf_t(B_{i,a}) + R(B_{i,a}) \Big)$$

$$\leq \Pr\Big(\big| \bar{s}_t(B_{i,a}) - \mu(a, O_{B_{i,a}}) \big| - \big| \mathbf{E}\left[\bar{s}_t(B_{i,a}) \right] - \mu(a, O_{B_{i,a}}) \big| > conf_t(B_{i,a}) \Big)$$

$$\leq \Pr\Big(\big| \bar{s}_t(B_{i,a}) - \mathbf{E}\left[\bar{s}_t(B_{i,a}) \right] \big| > conf_t(B_{i,a}) \Big)$$

$$\leq 2 \exp\left(-2n_t(B_{i,a}) conf_t(B_{i,a})^2 \right)$$

$$\leq 2 \exp\left(-\frac{2n_t(B_{i,a})}{n_t(B_{i,a}) + 1} \cdot 4 \log T \right)$$

$$\leq 2 T^{-4}$$

Since each agent will generate at most one new ball in each round, the total number of balls with $n_t(B_{i,a}) \geq 1$ and $R(B_{i,a}) < 1$ is at most nt for any t. To complete the proof, we apply the Union bound over all rounds t and all such balls B,

$$\Pr\left[\text{bad event} \right] \leq T \cdot nT \cdot 2T^{-4} = 2nT^{-2}$$

□

Then, we show that the global regret can be upper bounded by the sum of the confidence of the action taken up to a constant factor.

Lemma 2. *Assume Eqs. (1) to (3) hold. With probability at least $1 - 2nT^{-2}$, the global regret over T rounds is bounded by*

$$Reg_T \leq 2\alpha \sum_{t=1}^{T} \sum_{j=1}^{m} \sum_{i \in \mathcal{C}^j} U_t(a_{i,t})$$

Proof of Lemma 2. Consider the global regret incurred in any round t, by Lipschitz condition, we have

$$Reg_t = \mathcal{F}\left(\mathbf{a}_t^*, \boldsymbol{\mu}_t \right) - \mathcal{F}\left(\mathbf{a}_t, \boldsymbol{\mu}_t \right)$$

$$\leq \mathcal{F}\left(\mathbf{a}_t^*, \hat{\boldsymbol{\mu}}_t \right) - \mathcal{F}\left(\mathbf{a}_t, \boldsymbol{\mu}_t \right) + \alpha \sum_{j=1}^{m} \sum_{i \in \mathcal{C}^j} \left| \hat{\mu}(a_{i,t}^*, x_{i,t}) - \mu(a_{i,t}^*, x_{i,t}) \right|$$

Then according to the optimality of $\mathbf{a}_t$, we obtain

$$\mathcal{F}\left(\mathbf{a}_t^*, \hat{\boldsymbol{\mu}}_t \right) \leq \mathcal{F}\left(\mathbf{a}_t, \hat{\boldsymbol{\mu}}_t \right) + \alpha \sum_{j=1}^{m} \sum_{i \in \mathcal{C}^j} U_t(a_{i,t}) - \alpha \sum_{j=1}^{m} \sum_{i \in \mathcal{C}^j} U_t(a_{i,t}^*)$$

Combining these two inequalities and by Lipschitz condition, we get

$$Reg_t \leq \mathcal{F}(a_t, \hat{\mu}_t) - \mathcal{F}(a_t, \mu_t) + \alpha \sum_{j=1}^{m} \sum_{i \in \mathcal{C}^j} U_t(a_{i,t}) - \alpha \sum_{j=1}^{m} \sum_{i \in \mathcal{C}^j} U_t(a_{i,t}^*)$$

$$+ \alpha \sum_{j=1}^{m} \sum_{i \in \mathcal{C}^j} \left| \hat{\mu}(a_{i,t}^*, x_{i,t}) - \mu(a_{i,t}^*, x_{i,t}) \right|$$

$$\leq \alpha \sum_{j=1}^{m} \sum_{i \in \mathcal{C}^j} \left| \hat{\mu}(a_{i,t}, x_{i,t}) - \mu(a_{i,t}, x_{i,t}) \right| + \alpha \sum_{j=1}^{m} \sum_{i \in \mathcal{C}^j} U_t(a_{i,t}) - \alpha \sum_{j=1}^{m} \sum_{i \in \mathcal{C}^j} U_t(a_{i,t}^*)$$

$$+ \alpha \sum_{j=1}^{m} \sum_{i \in \mathcal{C}^j} \left| \hat{\mu}(a_{i,t}^*, x_{i,t}) - \mu(a_{i,t}^*, x_{i,t}) \right|$$

Now fix a single agent i. For simplicity, we might drop some subscript i in the subsequent equations, but implicitly all terms correspond to agent i. Consider the action $a_{i,t}$ taken and the corresponding ball $\hat{B}_{i,a_{i,t}}$ selected by agent i in round t. Denote the representative ball of $\hat{B}_{i,a_{i,t}}$ as $\hat{B}_{rep}$. By Lemma 1, under good event, we have

$$\left| \bar{s}_t(\hat{B}_{rep}) - \mu(a_{i,t}, O_{\hat{B}_{rep}}) \right| \leq conf_t(\hat{B}_{rep}) + R(\hat{B}_{rep})$$

By Lipschitz condition,

$$\left| \mu(a_{i,t}, O_{\hat{B}_{rep}}) - \mu(a_{i,t}, O_{\hat{B}_{i,a_{i,t}}}) \right| \leq D(\hat{B}_{i,a_{i,t}}, \hat{B}_{rep})$$

$$\left| \mu(a_{i,t}, O_{\hat{B}_{i,a_{i,t}}}) - \mu(a_{i,t}, x_{i,t}) \right| \leq R(\hat{B}_{i,a_{i,t}})$$

Combining above inequalities, we obtain

$$\left| \bar{s}_t(\hat{B}_{rep}) - \mu(a_{i,t}, x_{i,t}) \right| \leq conf_t(\hat{B}_{rep}) + D(\hat{B}_{i,a_{i,t}}, \hat{B}_{rep}) + R(\hat{B}_{rep}) + R(\hat{B}_{i,a_{i,t}})$$
$$= U_t(\hat{B}_{i,a_{i,t}}) = U_t(a_{i,t})$$

According to Eqs. (5) and (7), $\hat{\mu}_t(a_{i,t}, x_{i,t}) = s_t(\hat{B}_{i,a_{i,t}}) = \bar{s}_t(\hat{B}_{rep})$. Thus, we have

$$\left| \hat{\mu}_t(a_{i,t}, x_{i,t}) - \mu(a_{i,t}, x_{i,t}) \right| \leq U_t(a_{i,t})$$

Similarly, $|\hat{\mu}_t(a_{i,t}^*, x_{i,t}) - \mu(a_{i,t}^*, x_{i,t})| \leq U_t(a_{i,t}^*)$. Therefore, under good event,

$$Reg_t \leq 2\alpha \sum_{j=1}^{m} \sum_{i \in \mathcal{C}^j} U_t(a_{i,t})$$

The global regret is simply the sum of Reg_t in all T rounds.   □

Now let us consider a single agent i and establish an upper bound for $U_t(a_{i,t})$. There are two possibilities. (1) When the ball selected $\hat{B}_{i,a_{i,t}}$ is a parent ball; (2) When $\hat{B}_{i,a_{i,t}}$ is a non-parent ball;

Lemma 3. *Assume Eqs. (1) and (2) hold. With probability at least $1 - 2nT^{-2}$, for $\forall t \in [T], i \in \mathcal{C}$, $U_t(a_{i,t})$ is bounded by*

$$U_t(a_{i,t}) \le 7R(\hat{B}_{i,a_{i,t}})$$

Moreover, if $\hat{B}_{i,a_{i,t}}$ is a parent ball in round t, the bound can be improved to

$$U_t(a_{i,t}) \le 3R(\hat{B}_{i,a_{i,t}})$$

Proof of Lemma 3. We use $\hat{B}_{par}$ and $\hat{B}_{rep}$ to denote the parent ball and the representative of $\hat{B}_{i,a_{i,t}}$ respectively. Since $\hat{B}_{rep} = \arg\min_{B \in \mathcal{B}^t_{i,a_{i,t}}} D(\hat{B}_{i,a_{i,t}}, B) + conf_t(B) + R(B)$, we have

$$U_t(a_{i,t}) = U_t(\hat{B}_{i,a_{i,t}}) = D(\hat{B}_{i,a_{i,t}}, \hat{B}_{rep}) + conf_t(\hat{B}_{rep}) + R(\hat{B}_{rep}) + R(\hat{B}_{i,a_{i,t}})$$
$$\le D(\hat{B}_{i,a_{i,t}}, \hat{B}_{par}) + conf_t(\hat{B}_{par}) + R(\hat{B}_{par}) + R(\hat{B}_{i,a_{i,t}})$$

By the rule of parent ball, we have

$$conf_t(\hat{B}_{par}) \le R(\hat{B}_{par})$$
$$D(\hat{B}_{i,a_{i,t}}, \hat{B}_{par}) \le R(\hat{B}_{par})$$
$$R(\hat{B}_{par}) = 2R(\hat{B}_{i,a_{i,t}})$$

Therefore,

$$U_t(a_{i,t}) \le 3R(\hat{B}_{par}) + R(\hat{B}_{i,a_{i,t}}) = 7R(\hat{B}_{i,a_{i,t}})$$

In cases when $\hat{B}_{i,a_{i,t}}$ is a parent ball, similarly, by the rule of parent ball, we have

$$U_t(a_{i,t}) = D(\hat{B}_{i,a_{i,t}}, \hat{B}_{rep}) + conf_t(\hat{B}_{rep}) + R(\hat{B}_{rep}) + R(\hat{B}_{i,a_{i,t}})$$
$$\le D(\hat{B}_{i,a_{i,t}}, \hat{B}_{i,a}) + conf_t(\hat{B}_{i,a_{i,t}}) + R(\hat{B}_{i,a_{i,t}}) + R(\hat{B}_{i,a_{i,t}})$$
$$\le conf_t(\hat{B}_{i,a_{i,t}}) + R(\hat{B}_{i,a_{i,t}}) + R(\hat{B}_{i,a_{i,t}})$$
$$\le 3R(\hat{B}_{i,a_{i,t}})$$

$\square$

To continue with the proof, define $\mathcal{B}^T_{i,a}(r)$ as the collection of balls of radius r in $\mathcal{B}^T_{i,a}$, i.e.,

$$\mathcal{B}^T_{i,a}(r) = \{B \in \mathcal{B}^T_{i,a} | R(B) = r\}$$

Then, let's derive an upper bound for the number of balls with radius r in $\mathcal{B}^T_{i,a}$.

Lemma 4. *For $\forall i$, $\forall a \in \mathcal{A}_i$ and $\forall r = 2^{-k}, k \in \mathbb{N}$,*

$$\left| \mathcal{B}^T_{i,a}(r) \right| \le N_r(\mathcal{X}_i)$$

where $N_r(\mathcal{X}_i)$ is the r-covering number of $\mathcal{X}_i$

Proof of Lemma 4. First, for any $i \in \mathcal{C}$ and $a \in \mathcal{A}_i$, we show that in $\mathcal{B}_{i,a}^T$, the centers of balls whose radius is r are within distance at least r from one another. Consider $\forall B_{i,a}, B_{i,a}' \in \mathcal{B}_{i,a}^T$ where $R(B_{i,a}) = R(B_{i,a}') = r$ and $B_{i,a}$ and $B_{i,a}'$ are generated at round t and t' respectively. Without loss of generality, assume that $t < t'$. Let B_{par}' denote the parent of $B_{i,a}'$. Recall that $\mathrm{dom}_t(B_{par}')$ is a subset of B_{par}' that excludes all balls in $\mathcal{B}_{i,a}^t$ with a smaller radius. Thus, $\mathrm{dom}_{t'}(B_{par}') \cap B_{i,a} = \varnothing$. Moreover, according to the rule of parent, $O_{B_{i,a}'} \in \mathrm{dom}_{t'}(B_{par}')$. As a result, $O_{B_{i,a}'} \notin B_{i,a}$. and therefore we have

$$D(B_{i,a}, B_{i,a}') \geq r$$

Now we proceed with the proof. Suppose $\left|\mathcal{B}_{i,a}^T(r)\right| = N_r(\mathcal{X}_i) + 1$, which means that there are $N_r(\mathcal{X}_i) + 1$ balls $B_1, \ldots, B_{N_r(\mathcal{X}_i)+1}$ of radius r in $\mathcal{B}_{i,a}^T$. Then by pigeonhole, we must have two balls B_m and B_n whose centers fall into the same $B(v_h, \frac{r}{2})$. This means that the distance between the centers of these two balls cannot be more than the diameter of the ball $B(v_h, \frac{r}{2})$, i.e., $D(B_m, B_n) \leq r$, which leads to a contradiction. Therefore, we have

$$\left|\mathcal{B}_{i,a}^T(r)\right| \leq N_r(\mathcal{X}_i)$$

$\square$

Now we are ready to prove the theorem.

Proof of Theorem 1. Let $Reg_T^i = 2\alpha \sum_{t=1}^T U_t(a_{i,t})$. Then we can separate the global regret into individual ones, i.e., $Reg_T \leq \sum_{j=1}^m \sum_{i \in \mathcal{C}^j} Reg_T^i$. Now, consider a single agent i. For any ball $B \in \mathcal{B}_i$, let A_B be a singleton set containing the round when B is activated and S_B denotes the set of rounds when B is selected and is not a parent ball. Then by construction, $\bigcup_{B \in \mathcal{B}_i} \{A_B, S_B\}$ forms a partition of set $[T]$. Moreover, note that in the round t when B is activated, the ball selected $\hat{B}_{i,a}$ must be a parent ball. Thus, if we use $\mathbb{1}\{\cdot\}$ to denote the indicator function, Reg_T^i can be represented as

$$Reg_T^i = 2\alpha \sum_{t \in [T]} \sum_{r \in \mathbf{R}(0,1)} \sum_{B \in \mathcal{B}_i^T(r)} \mathbb{1}\{t \in A_B \cup S_B\} \cdot U_t(a_{i,t})$$

$$\underbrace{= 2\alpha \sum_{t \in [T]} \sum_{r \in \mathbf{R}(0,r_i')} \sum_{B \in \mathcal{B}_i^T(r)} \mathbb{1}\{t \in A_B \cup S_B\} \cdot U_t(a_{i,t})}_{(1)}$$

$$\underbrace{+ 2\alpha \sum_{t \in [T]} \sum_{r \in \mathbf{R}(r_i',1)} \sum_{B \in \mathcal{B}_i^T(r)} \mathbb{1}\{t \in A_B \cup S_B\} \cdot U_t(a_{i,t})}_{(2)}$$

where r_i' can take any value in $(0,1)$. On one hand, by Lemma 3, we can obtain a bound for part (1):

$$(1) \leq \sum_{t \in [T]} \sum_{r \in \mathbf{R}(0,r_i')} \sum_{B \in \mathcal{B}_i^T(r)} \mathbb{1}\{t \in A_B \cup S_B\} \cdot 7r_i' \leq 7r_i'T$$

On the other hand, for part (2):

$$(2) = \sum_{t\in[T]} \sum_{r\in \mathbf{R}(r_i',1)} \sum_{B\in \mathcal{B}_i^T(r)} \mathbb{1}\{t \in A_B\} \cdot U_t(a_{i,t}) + \mathbb{1}\{t \in S_B\} \cdot U_t(a_{i,t})$$

$$\leq \sum_{t\in[T]} \sum_{r\in \mathbf{R}(r_i',1)} \sum_{B\in \mathcal{B}_i^T(r)} \mathbb{1}\{t \in A_B\} \cdot 6R(B) + \mathbb{1}\{t \in S_B\} \cdot 7R(B)$$

$$\leq \sum_{r\in \mathbf{R}(r_i',1)} \sum_{B\in \mathcal{B}_i^T(r)} 6r + 7r \cdot \left(\frac{4\log T}{r^2} - 2\right)$$

For the first inequality, when $t \in |S_B|$, B is the ball selected. Then by Lemma 2 and Lemma 3, we have $U_t(a_{i,t}) \leq 7R(B)$. On the other hand, $t \in A_B$ means that B is activated in time t. Then the ball selected $\hat{B}_{i,a_{i,t}}$ is the parent ball of B. Thus, we have $U_t(a_{i,t}) \leq 3R(\hat{B}_{i,a_{i,t}}) = 6R(B)$. For the second inequality, as defined, B is not a parent ball when $t \in A_B \cup S_B$. Thus, by the rule of parent, we can get an upper bound for the cardinality of $|S_B|$,

$$conf_t(B) = \sqrt{\frac{4\log T}{1 + n_t(B)}} > R(B)$$

It means that

$$n_t(B) = \sum_{t\in[T]} \mathbb{1}\{t \in A_B\} + \mathbb{1}\{t \in S_B\} < \frac{4\log T}{R(B)^2} - 1$$

Therefore, we have $|A_B| = 1$ and $|S_B| < \frac{4\log T}{R(B)^2} - 2$.

In addition, based on Lemma 4 and $\mathcal{B}_i = \bigcup_{a\in\mathcal{A}_i} \mathcal{B}_{i,a}$, we have $\left|\mathcal{B}_i^T(r)\right| \leq |\mathcal{A}_i| N_r(\mathcal{X}_i)$. Now we can bound the regret of agent i as follows:

$$(2) \leq \sum_{r\in \mathbf{R}(r_i',1)} \sum_{B\in \mathcal{B}_i^T(r)} 6r + 7r \cdot \left(\frac{4\log T}{r^2} - 2\right)$$

$$\leq |\mathcal{A}_i| \sum_{r\in \mathbf{R}(r_i',1)} 6N_r(\mathcal{X}_i)r + 7N_r(\mathcal{X}_i)r \cdot \left(\frac{4\log T}{r^2} - 2\right)$$

$$= |\mathcal{A}_i| \sum_{r\in \mathbf{R}(r_i',1)} \frac{28N_r(\mathcal{X}_i)\log T}{r} - 8N_r(\mathcal{X}_i)r$$

Combining (1) and (2), we get

$$Reg_T \leq 2\alpha \sum_{j=1}^{m} \sum_{i\in\mathcal{C}^j} \left(7r_i'T + |\mathcal{A}_i| \sum_{r\in \mathbf{R}(r_i',1)} \frac{28N_r(\mathcal{X}_i)\log T}{r} - 8N_r(\mathcal{X}_i)r\right).$$

Therefore, we complete the proof.

$$Reg_T \leq 2\alpha \sum_{j=1}^{m} \sum_{i\in\mathcal{C}^j} \inf_{r_i'\in(0,1)} \left(7r_i'T + |\mathcal{A}_i| \sum_{r\in \mathbf{R}(r_i',1)} \frac{28N_r(\mathcal{X}_i)\log T}{r}\right)$$

$\square$

Moreover, define the **covering dimension** d_i for any agent i as

$$d_i \triangleq \inf \left\{ d > 0 : N_r(\mathcal{X}_i) \leq \beta r^{-d_i} \quad \forall r \in (0,1) \right\}.$$

Then substitute $N_r(\mathcal{X}_i) \leq \beta r^{-d_i}$ into the inequality in Theorem 1, we obtain the following corollary.

Corollary 1.

$$Reg_T \leq O(KT^{\frac{d+1}{d+2}} \log T^{\frac{1}{d+2}}) \leq \tilde{O}(T^{\frac{d+1}{d+2}})$$

where $d = \max(d_1, \ldots, d_n)$ *and* $K = \sum_{j=1}^{m} \sum_{i \in \mathcal{C}^j} |\mathcal{A}_i|$.

In addition, the sublinear regret bounds of MASCUCB and MASCUCBwF can be derived in a similar way to Theroem 1.

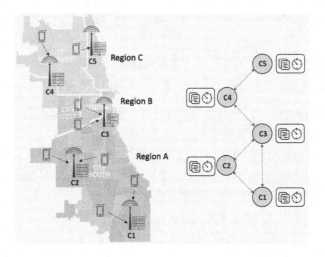

Fig. 3. Loose couplings and context information

6   Experiment

In this section, we evaluate the performance of the proposed algorithms in a real-life scenario.

6.1   Experiment Setting

The dataset used is the AuverGrid dataset from the Grid Workloads Archive (GWA) [11]. This dataset records the real-word computational demand received by large-scale multi-site infrastructures to support e-Science. It contains 400k task requests of 5 grids.

The learning problem is formulated as follows. Consider each cloudlet as an agent. There are 5 cloudlets in total. As shown in Fig. 3, cloudlets [C1, C2, C3] serve users in region A, [C3, C4] provide services in region B and region C has two cloudlets [C4, C5]. In each round, side-information is observed by each cloudlet. Then rental decisions need to be determined. There are some options available, i.e., $\mathcal{A} = [0, 2, 4, 6]$, corresponding to the number of virtual machines (VM) to rent by each cloudlet. The goal is to maximize ASP's global utility.

Table 1. Hyperparameter

Parameter	Value
Time horizon T	5000
2D Context space $Context$	[Demand in last 24 h, Current time]
Input data size per task S	1 MB
Required CPU cycles per task C	10^9
CPU frequency of each VM F	2×10^9 Herts
Price per VM P	0.1 unit
Maximum service demand per VM D_{max}	80
Expected transmission rate of cloudlets R_c	5 Mbps
Expected transmission rate of the cloud R_{remote}	2 Mbps
Expected backbone transmission rate R_b	10 Mbps
Processor capacity per task at Cloud center V	10×10^9 Herts
Round-trip travel time to the Cloud h_t	1

More specifically, the score function measures the quality of service (QoS) minus the cost incurred. Herein QoS is measured as the processing time saved by computing at cloudlets instead of the remote Cloud. For each task, the processing time per task at cloudlets equals to transmission delay plus processing delay. Take Cloudlet C1 as an example. If the number of VMs rented at C1 is a_1 with $a_1 > 0$, then the processing time at C1 is $T_1 = \frac{S}{R_c} + \frac{C}{Fa_1}$. In comparison, the processing time at the remote Cloud is $T_{Remote} = \frac{S}{R_{\text{remote}}} + \frac{S}{R_b} + \frac{C}{V} + h_t$. Thus, QoS per task at C1 is $QoS = \Delta t = T_{Remote} - T_1$. In cases when $a_1 = 0$, QoS is set to zero. While the rental cost of VMs is $Cost_1 = Pa_1$. Let d_t be the service demand (number of tasks) covered by C1. Then, given the maximum service demand processed per VM D_{max}, the score achieved by taking action a_1 at C1 is $\mathcal{S}_1 = min(d_t, a_1 D_{max})QoS - Cost_1$. Similar formulation has been considered in [6].

The joint utility in each region can take any form as long as it satisfies Eqs. (2) and (3). In this experiment, we use the Worst Performance Metric to measure the local utility in one region, i.e. $f_a = min\big(s_1(a_1, x_1), s_2(a_2, x_2), s_3(a_3, x_3)\big)$. Meanwhile, the global utility is the sum of the utilities in all regions $\mathcal{F} = f_a + f_b + f_c$. Please see Table 1 for details about the parameter configurations.

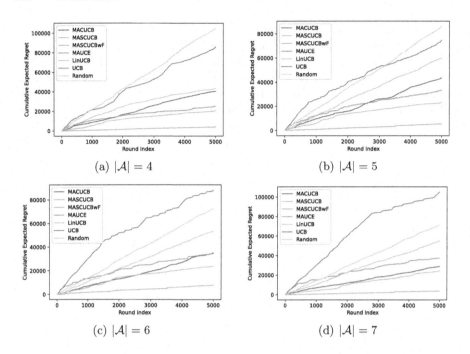

(a) $|\mathcal{A}| = 4$ (b) $|\mathcal{A}| = 5$

(c) $|\mathcal{A}| = 6$ (d) $|\mathcal{A}| = 7$

Fig. 4. Comparison of cumulative regrets between different algorithms

6.2   Experimental Results

To test its performance, we compared MACUCB algorithm and its variants with some classical algorithms:

UCB1 [2]: UCB1 can also be applied to multi-agent cases. The key idea is to treat each possible combination of rental decisions as a different action.

LinUCB [14]: LinUCB makes use of the context information of agents by assuming that the expected utility is defined by the inner product between the context vector and an unknown coefficient vector.

MAUCE [3]: MAUCE is an algorithm for multi-agent multi-armed bandits, which also exploits loose couplings.

MASCUCB: In MASCUCB, agents share the collection of balls for estimation. By sharing historical observations, agents could take advantage of the similarities to make better decisions.

MASCUCBwF: For computing resource rental problem, the scores of other actions are also revealed once we observe the service demand. Thus, MAS-CUCBwF makes full use of this extra information to adjust the estimation.

Random: The algorithm randomly selects a possible combination of resource rental decisions in each round.

Figure 4 depicts the cumulative regrets incurred by these algorithms under different rental option sets in the first 5000 rounds. It can be seen that MACUCB and its variants significantly outperform other benchmarks across the time period considered when the action set is large. Specifically, the rental options selected to conduct the experiments are $\mathcal{A} = [0, 2, 4, 6]$, $\mathcal{A} = [0, 2, 4, 6, 8]$, $\mathcal{A} = [0, 2, 4, 6, 8, 10]$ and $\mathcal{A} = [0, 2, 4, 6, 8, 10, 12]$ respectively.

Note that the performance of UCB1 degrades significantly with increasing action set. It is even worse than Random algorithm when $|\mathcal{A}| \geq 6$. This poor performance can be mainly explained by two reasons: Firstly, since UCB1 treats each possible combination of rental decisions as an action, the problem size grows exponentially with the number of agents. As a result, UCB1 needs to spend a large fraction of time in the exploration phase, making it inefficient. Secondly, UCB1 fails to establish a link between contexts and utilities. Although LinUCB considers the context information in the estimation, its performances are worse than MAUCE, MACUCB and its variants. Same as UCB1, due to a large joint action set, LinUCB spends too much time in the exploration phase, preventing it from taking the optimal action frequently. Similar to our algorithm, MAUCE also exploits loose couplings. Although it achieves smaller regrets than MACUCB when the size of the action set is small, the gap narrows with increasing action set. Eventually, MACUCB outperforms MAUCE when $|\mathcal{A}| = 7$. Moreover, since MAUCE fails to exploit the similarities across agents, it performs worse than MASCUCB and MASCUCBwF across all sessions. In addition, it is highly helpful to exploit the similarities between cloudlets, as evidenced by comparing the performance of MASCUCB and MACUCB. Figure 4 shows that MASCUCB outperforms MACUCB in all sessions. Furthermore, comparing MASCUCB and MASCUCBwF, we see that observing more information about scores of actions at cloudlets increases the accuracy of estimation and results in a lower cumulative regret. Since we are using historical data to estimate current scores, uncertainty is always present. The Lipschitz assumption between context and scores only roughly holds. Having more information about real scores of actions could help to correct the bias and reduce the uncertainty to some extent.

7   Conclusion

In this paper, we formulate the multi-agent coordination problem as a multi-agent contextual bandit problem and an online algorithm called MACUCB is proposed to address it. To efficiently perform the maximization in multi-agent settings, MACUCB applies a variable elimination technique to exploit loose couplings. Meanwhile, a modified zooming technique is used in MACUCB to adaptively exploit the context information. Besides, two enhancement methods are proposed which achieve better theoretical and practical performance. One shares the common context space among the agents and the other makes use of full feedback information available. Moreover, sublinear regret bounds were derived for each of the proposed algorithms. Finally, the experiment results on a real-world dataset show that the proposed algorithms outperform other benchmarks.

References

1. Audibert, J.Y., Bubeck, S., Lugosi, G.: Minimax policies for combinatorial prediction games. In: Proceedings of the 24th Annual Conference on Learning Theory, pp. 107–132 (2011)
2. Auer, P., Cesa-Bianchi, N., Fischer, P.: Finite-time analysis of the multiarmed bandit problem. Mach. Learn. **47**(2–3), 235–256 (2002)
3. Bargiacchi, E., Verstraeten, T., Roijers, D., Nowé, A., Hasselt, H.: Learning to coordinate with coordination graphs in repeated single-stage multi-agent decision problems. In: International Conference on Machine Learning, pp. 491–499 (2018)
4. Bubeck, S., Cesa-Bianchi, N., et al.: Regret analysis of stochastic and nonstochastic multi-armed bandit problems. Found. Trends® Mach. Learn. **5**(1), 1–122 (2012)
5. Cesa-Bianchi, N., Lugosi, G.: Combinatorial bandits. J. Comput. Syst. Sci. **78**(5), 1404–1422 (2012)
6. Chen, L., Xu, J.: Budget-constrained edge service provisioning with demand estimation via bandit learning. IEEE J. Sel. Areas Commun. **37**(10), 2364–2376 (2019)
7. Chen, W., Wang, Y., Yuan, Y.: Combinatorial multi-armed bandit: general framework and applications. In: International Conference on Machine Learning, pp. 151–159 (2013)
8. De, Y.M., Vrancx, P., Nowé, A.: Learning multi-agent state space representations. In: Proceedings of 9th International Conference of Autonomous Agents and Multiagent Systems, pp. 715–722 (2010)
9. Gai, Y., Krishnamachari, B., Jain, R.: Combinatorial network optimization with unknown variables: multi-armed bandits with linear rewards and individual observations. IEEE/ACM Trans. Network. **20**(5), 1466–1478 (2012)
10. Guestrin, C., Koller, D., Parr, R.: Multiagent planning with factored MDPS. In: Advances in Neural Information Processing Systems, pp. 1523–1530 (2002)
11. Iosup, A., et al.: The grid workloads archive. Fut. Gener. Comput. Syst. **24**(7), 672–686 (2008)
12. Kok, J.R., Spaan, M.T., Vlassis, N., et al.: Multi-robot decision making using coordination graphs. In: Proceedings of the 11th International Conference on Advanced Robotics, ICAR, vol. 3, pp. 1124–1129 (2003)
13. Kok, J.R., Vlassis, N.: Collaborative multiagent reinforcement learning by payoff propagation. J. Mach. Learn. Res. **7**, 1789–1828 (2006)
14. Li, L., Chu, W., Langford, J., Schapire, R.E.: A contextual-bandit approach to personalized news article recommendation. In: Proceedings of the 19th International Conference on World Wide Web, pp. 661–670 (2010)
15. Qin, L., Chen, S., Zhu, X.: Contextual combinatorial bandit and its application on diversified online recommendation. In: Proceedings of the 2014 SIAM International Conference on Data Mining, pp. 461–469. SIAM (2014)
16. Roijers, D.M., Whiteson, S., Oliehoek, F.A.: Computing convex coverage sets for faster multi-objective coordination. J. Artif. Intell. Res. **52**, 399–443 (2015)
17. Rollón, E., Larrosa, J.: Bucket elimination for multiobjective optimization problems. J. Heurist. **12**(4–5), 307–328 (2006)
18. Scharpff, J., Roijers, D.M., Oliehoek, F.A., Spaan, M.T., de Weerdt, M.M.: Solving transition-independent multi-agent MDPS with sparse interactions. In: Thirtieth AAAI Conference on Artificial Intelligence, pp. 3174–3180 (2016)
19. Scharpff, J., Spaan, M.T., Volker, L., De Weerdt, M.M.: Planning under uncertainty for coordinating infrastructural maintenance. In: Twenty-Third International Conference on Automated Planning and Scheduling, pp. 169–170 (2013)

20. Slivkins, A.: Contextual bandits with similarity information. J. Mach. Learn. Res. **15**(1), 2533–2568 (2014)
21. Verstraeten, T., Bargiacchi, E., Libin, P.J., Helsen, J., Roijers, D.M., Nowé, A.: Thompson sampling for loosely-coupled multi-agent systems: An application to wind farm control. In: Adaptive and Learning Agents Workshop 2020, ALA 2020 (2020). https://ala2020.vub.ac.be
22. Wiering, M.: Multi-agent reinforcement learning for traffic light control. In: Machine Learning: Proceedings of the Seventeenth International Conference (ICML 2000), pp. 1151–1158 (2000)

Battery Management for Automated Warehouses via Deep Reinforcement Learning

Yanchen Deng[1(✉)], Bo An[1], Zongmin Qiu[2], Liuxi Li[2], Yong Wang[2], and Yinghui Xu[2]

[1] School of Computer Science and Engineering, Nanyang Technological University, Singapore, Singapore
{ycdeng,boan}@ntu.edu.sg
[2] Cainiao Smart Logistics Network, Hangzhou, China
{zongmin.qzm,liuxi.llx,richard.wangy}@cainiao.com, renji.xyh@taobao.com

Abstract. Automated warehouses are widely deployed in large-scale distribution centers due to their ability of reducing operational cost and improving throughput capacity. In an automated warehouse, orders are fulfilled by battery-powered AGVs transporting movable shelves or boxes. Therefore, battery management is crucial to the productivity since recovering depleted batteries can be time-consuming and seriously affect the overall performance of the system by reducing the number of available robots. In this paper, we propose to solve the battery management problem by using deep reinforcement learning (DRL). We first formulate the battery management problem as a Markov Decision Process (MDP). Then we show the state-of-the-art DRL method which uses Gaussian noise to enforce exploration could perform poorly in the formulated MDP, and present a novel algorithm called TD3-ARL that performs effective exploration by regulating the magnitude of the outputted action. Finally, extensive empirical evaluations confirm the superiority of our algorithm over the state-of-the-art and the rule-based policies.

Keywords: Automated warehouses · Battery management · Deep reinforcement learning

1 Introduction

With the rapid development of multi-robot systems, automated warehouses and Robotic Mobile Fulfillment Systems (RMFSs), such as the Kiva system [4], have emerged as a new category of automated order fulfillment systems. In such systems, order fulfillment is implemented by a fleet of battery-powered Automatic Guided Vehicles (AGVs) to transport movable shelves with Stock-keeping Units (SKUs). Due to their ability of reducing operational cost and improving throughput capacity, automated warehouses have been adapted in many e-commerce companies including Amazon, Cainiao, etc. Battery management is crucial to

© Springer Nature Switzerland AG 2020
M. E. Taylor et al. (Eds.): DAI 2020, LNAI 12547, pp. 126–139, 2020.
https://doi.org/10.1007/978-3-030-64096-5_9

automated warehouses as recovering the depleted batteries is time-consuming and significantly affects the throughput capacity of the system by reducing the number of available AGVs. Moreover, since the number of charging poles is fixed, inappropriate battery management would result in many low-power AGVs which cannot be recharged timely. As a result, these AGVs eventually cannot fulfill any job (including recharging) due to the extremely low battery and must be recovered manually, which makes battery management quite challenging.

A common way to implement battery management is by hand-crafted rules. For example, we could use a parameterized policy with two thresholds $\langle T_c, T_w \rangle$. The AGVs with battery lower than T_c are scheduled to charge and the AGVs with battery higher than T_w are scheduled to work. However, since most of them are built upon energy conservation, the rule-based policies are decoupled from the current workload and may schedule many AGVs to charge even during peak periods. As a result, the backlogs increase and the performance of the system degenerates significantly due to the lack of available AGVs. In other words, the rule-based policies are pre-defined and cannot be adjusted adaptively according to the real-time workload.

To maximize throughput capacity, it is necessary to explore alternative approaches to solve the battery management problem by explicitly considering the dynamic workload. Our first contribution is modeling the battery management problem as a Markov Decision Process (MDP) due to its capability in modeling long-term planning problems with uncertainty. The state of the formulated MDP includes the battery histograms of AGVs under different states, the number of working AGVs in different working areas and the number of backlogs. The action is defined as the upper bound of the AGVs in the working areas and the threshold for the charging AGVs. The reward is defined as the number of fulfilled orders in each time step.

Solving the MDP is challenging since both the state space and the action space are continuous. Traditional tabular-based reinforcement learning approaches [2,8,15,20] cannot be applied due to their inability of modeling high-dimensional and complicated dynamics. Pioneered by DQN [13], deep reinforcement learning (DRL) [5] has achieved tremendous success in solving many sequential decision-making problems [14,16,18]. TD3 [6] is the state-of-the-art DRL algorithm for continuous control. However, it implements exploration by adding Gaussian noise to the outputted action, which could be inefficient in our problem where the permuted action may still has the same semantic if the magnitude of the outputted action is high. Therefore, our second contribution is proposing a novel algorithm called TD3 with action regulation loss (TD3-ARL) to enforce the state dependent exploration. In more detail, we regulate the magnitude of the outputted action by imposing a loss term on the objective function of the actor network to guarantee the diversity of exploration. Our extensive empirical evaluations have demonstrated the superiority of TD3-ARL over the state-of-the-art.

2   Related Work

Battery Management for Automated Warehouses. While battery management is crucial to large-scale automated warehouses, its influence on the performance is usually omitted in automated warehouse studies [9]. McHANEY examined several charging schemes and pointed out the battery constraint can only be omitted when charging can be insured to take place without impacting system operation [11]. Recently, Zou et al. evaluated the performance of different recovering strategies including re-charging, swapping and inductive charging [22]. Several heuristics for dispatching low-power AGVs for battery swapping were proposed in [3]. However, these heuristics cannot be applied to our case since we focus on recovering depleted batteries via re-charging.

Deep Reinforcement Learning. Combining with high-capacity deep neural network approximators, DRL [5] has achieved great successes on challenging decision making problems [14,16,18]. Pioneered by DQN [13], many DRL algorithms such as A3C [12], DDPG [10], SAC [7] and TD3 [6] have been proposed. However, most of off-policy DRL algorithms use uncorrelated Gaussian noise to enforce exploration, which would perform poorly in our case where many actions are equivalent. Therefore, based on TD3 we propose a novel algorithm which performs effective exploration by regulating the magnitudes of the outputted actions.

3   Motivation Scenario

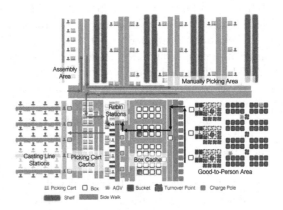

Fig. 1. The layout of a typical automated warehouse

Figure 1 gives the layout of a typical automated warehouse. Generally, the life-cycle of an order includes picking, consolidation and casting line. When an order comes in, the system assigns picking jobs for Good-to-Person Area to pick the required SKUs from buckets and store them to a turnover box. After finishing picking jobs, the box is sent to the Rebin Area for further consolidation. If the order does not contain any manually picking item, the system assigns an assembly job for Assembly Area to assemble a picking-cart with empty parcels. Otherwise, the system assigns an assembly job for the Manually Picking Area to pick the SKUs from the shelves manually and assemble a picking-cart in which parcels are filled with the required SKUs. The picking-cart is sent to the Rebin Area as soon as the assembly job finishes. After both the turnover box and the picking-cart arrived at the Rebin Area, the consolidation is initiated to combine the SKUs from different areas by transferring SKUs in the box to the parcels in the picking-cart. After finishing consolidation job, the picking-cart is sent to the Casting Line Area for packaging and inspecting. The order is considered as fulfilled and sent outbound after the casting line job is finished.

Since the number of charging poles are fixed and all the transportation of buckets, boxes and picking-carts is fulfilled by battery-powered AGVs, inappropriate battery management would lead to the shortage of available AGVs and severely degenerates the throughput capacity of warehouses. Typically, battery management is implemented by thresholds. That is, the system schedules the AGVs with battery lower than a threshold to charge and schedules the charging AGVs with battery higher than another threshold to work. Besides, a working AGV will not execute charging job if the number of charging AGVs has already met the number of charging poles. Finally, an emergency charging mechanism that allows low-power AGVs to preempt charging poles is introduced to ensure that the battery of an AGV is higher than a safe threshold.

Although the rule-based policy is easy to be implemented and highly interpretable, it suffers from three major shortcomings. First, AGVs are scheduled to charge as long as their battery is lower than the charging threshold and there are available charging poles. As a result, it is possible to schedule a large number of AGVs to charge and cause the shortage of available AGVs, which is not desirable during peak periods. Second, the thresholds are decoupled from the current workload, which prohibits the system from improving throughput capacity proactively when needed. In fact, a common way to improve productivity in practice is to disable the charging poles manually to force AGVs to work for a longer time when the working areas are busy. However, this workaround is risky since reducing the number of charging poles would lead to a large number of low-power AGVs or even dead AGVs. Finally, the rule-based policy fails to exploit order structure. Specifically, since consolidation and casting line depend on picking, it is reasonable to schedule AGVs in picking areas to charge when most orders are in consolidation or casting line procedure. Given these limitations, we aim to take the current workload and the order structure into the consideration and propose a novel battery management scheme to maximize overall throughput capacity.

4   Problem Statement and MDP Formulation

In this section, we give the formal definition to battery management problem and formulate it as an MDP.

4.1   Battery Management Problem

We consider an automated warehouse with a fleet of AGVs G which are subject to the battery constraint, i.e., an AGV $g \in G$ cannot fulfill any job if its battery b_g is lower than the dead threshold T_{dead}. Orders arrive over time and each order includes picking jobs for Good-to-Person area, an assembly job for either Assembly area or Manual Picking area, and jobs for Rebin area and Cast Line area respectively. Given the number of charging poles $C < |G|$, the goal is to design a battery management scheme to maximize overall throughput capacity. More specifically, at each time step t, we aim to determine the set of charging AGVs $G_c^t \subset G$ and the set of AGVs for working areas $G_w^t \subseteq G$ such that $|G_c^t| \leq C, G_c^t \cap G_w^t = \emptyset$ and $G_c^t \cup G_w^t = G$ to maximize the number of fulfilled orders by the end of a day.

4.2   MDP Formulation

We propose to model the problem as a Markov Decision Process (MDP) due to its ability in modeling sequential decision-making problems with uncertainty. Formally, the MDP is defined by a tuple $\mathcal{M} = \langle \mathcal{S}, \mathcal{A}, \mathcal{R}, \mathcal{P} \rangle$ where $\mathcal{S}, \mathcal{A}, \mathcal{R}, \mathcal{P}$ are the state space, the action space, the reward function and the transition probability function, respectively. The definitions are given as follows.

- **State** $s^t \in \mathcal{S}$. The state we consider includes battery feature, AGV fleet feature and system feature. For the battery feature, we build a histogram with a bin size 5% for AGVs in charging state, working state and idle state, respectively. For the AGV fleet feature, we consider the number of working AGVs in each area. For the system feature, we consider the number of backlogs and the current time step t, where backlogs are the unfinished orders by the current time step.
- **Action** $a^t \in \mathcal{A}$. It is impossible to directly determine G_w^t and G_c^t as dividing a set into two disjoint subsets would result in a prohibitive large action space. Instead, we consider to schedule *anonymously*. Formally, a^t is defined by a tuple $\langle U_w^t, T_c^t \rangle$ where $|G| - C \leq U_w^t \leq |G|$ is the upper bound of the number of AGVs in working areas and T_c^t is the battery threshold for the charging AGVs. When realizing a^t, the system schedules $\max(|G_w^t| - U_w^t, 0)$ AGVs with the lowest battery to charge and schedules the charging AGVs with battery higher than T_c^t to working areas.
- **Reward function** $\mathcal{R}$. The reward is defined as the number of orders fulfilled in the time step t. We aim to find the policy $\pi^* : \mathcal{S} \to \mathcal{A}$ which maximizes the accumulated reward (i.e., the total number of fulfilled orders). That is,

$$\pi^* = \arg\max_\pi \sum_{t=1}^{T} \mathcal{R}(s^t, \pi(s^t))$$

5 Solving the MDP

Traditional RL algorithms cannot be applied to solve the MDP as the state space $\mathcal{S}$ and the action space $\mathcal{A}$ are continuous and the transition probability function $\mathcal{P}$ does not have an explicit form. Therefore, we resort to deep RL and present a novel algorithm built upon TD3. We will first briefly introduce TD3 and show it would perform poorly in our problem due to inefficient exploration. Then we present our proposed algorithm.

5.1 TD3

Twin Delayed Deep Deterministic Policy Gradient (TD3) [6] is the state-of-the-art deep reinforcement learning algorithm for continuous control. TD3 is a deterministic policy gradient algorithm [17] and incorporates an actor-critic architecture where both the actor and the critics are parameterized by deep neural networks. The policy which is represented by the actor network π_ϕ is updated to maximize

$$J(\phi) = \frac{1}{N} \sum_s Q_{\theta_1}(s, \pi_\phi(s)) \tag{1}$$

where N is the size of a mini-batch of experiences and Q_{θ_1} is a critic. To address the overestimation bias in critic learning, TD3 concurrently learns two critics $Q_{\theta_1}, Q_{\theta_2}$ and uses the smaller one to compute the targets for critic updates. The critics are updated to minimize the temporal difference error [19]. Besides, it adds noises to the target actions and enforces a delayed policy update to reduce variances in actor-critic methods.

Unfortunately, TD3 would perform poorly when solving MDP $\mathcal{M}$ due to inefficient exploration. Since TD3 trains a deterministic policy in an off-policy way, a common approach to enforce exploration is to add uncorrelated mean-zero Gaussian noise $\mathcal{N}(0, \sigma)$ to the outputted action, which would perform poorly in our case. In fact, many actions in MDP $\mathcal{M}$ are equivalent and the conventional exploration scheme fails to exploit the fact effectively. Consider the conventional exploration scheme applied to U_w^t shown in Fig. 2(a) where $\hat{U}_w^t$ is the deterministic action outputted by the actor. Since the system schedules $\max(|G_w^t| - U_w^t, 0)$ AGVs to charge, any $U_w^t \in [|G_w^t|, |G|]$ has the same semantic. As a result, TD3 could sample a lot of equivalent actions but rarely explore the region $[|G| - C, |G_w^t|]$ if the magnitude of the outputted action is close to $|G|$.

5.2 Enforcing State Dependent Exploration via Action Regulation Loss

Since all the actions in the range of $[|G_w^t|, |G|]$ are equivalent, we consider to tamp the magnitude of the outputted action such that $\hat{U}_w^t \leq |G_w^t| + \epsilon$ to ensure the

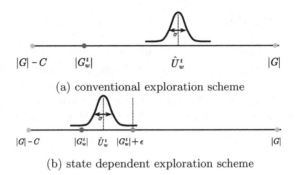

(a) conventional exploration scheme

(b) state dependent exploration scheme

Fig. 2. Different exploration schemes

diversity, where ϵ is a constant to achieve effective exploration (i.e., Fig. 2(b)). Thus, the bound is dependent on the current state. A straightforward way would be rounding the outputted action to the bound directly. However, as shown in [1], directly clipping would incur a bias on the policy gradient. Another approach to enforce the bound is reward engineering [21], which uses L_2-norm of the outputted action as a penalty and injects it directly into the reward function. However, since the rewards in $\mathcal{M}$ are objective quantities with real-world interpretation (i.e., the number of orders fulfilled in each time step), it is inappropriate to impose a hand-crafted penalty term. Besides, directly injecting the penalties to the rewards can cause high variances in critic learning since the critics bootstrap both the actual rewards and the penalties. In fact, the penalties do not have a temporal structure and should not be bootstrapped since the out-of-bound actions in different time steps are independent from each other.

Instead, we tamp the magnitude of the outputted action by directly imposing a loss term on the objective function of the actor and refer it as TD3 with action regulation loss (TD3-ARL). Formally, the actor network π_ϕ in our algorithm is updated to maximize

$$J(\phi) = \frac{1}{N} \sum_s Q_{\theta_1}(s,a) - L(s,a)|_{a=\pi_\phi(s)}, \tag{2}$$

where $L(s,a)$ is the regulate loss defined by

$$L(s,a) = |G_w(s) + \epsilon - U(a)| + U(a) - G_w(s) - \epsilon. \tag{3}$$

Here, $G_w : \mathcal{S} \to \mathbb{N}$ is a function that returns the number of AGVs in the working areas and $U : \mathcal{A} \to \mathbb{N}$ is a function that returns the first component of an action (i.e., the upper bound of the number of AGVs in working areas). When $U(a)$ is higher than the state dependent bound $G_w(s) + \epsilon$, a loss of $2(U(a) - G_w(s) - \epsilon)$ is incurred to regulate the magnitude of the action. Otherwise the loss term cancels and the objective function is equivalent to the one in vanilla TD3. Combining with uncorrelated Gaussian noises $\mathcal{N}(0,\sigma)$, the regulated action can achieve effective state dependent exploration. Algorithm 1 presents the sketch of TD3-ARL.

Algorithm 1: TD3 with Action Regulation Loss

Initialize critic networks $Q_{\theta_1}, Q_{\theta_2}$ and actor network π_ϕ with random parameters θ_1, θ_2 and ϕ
Initialize target networks $Q_{\theta_1'}, Q_{\theta_2'}, \pi_{\phi'}$ with $\theta_1' \leftarrow \theta_1, \theta_2' \leftarrow \theta_2, \phi' \leftarrow \phi$
Initialize replay memory M
for *episode=1,..., E* **do**
  Reset the environment and get the initial state s
  for $t=1,\ldots,T$ **do**
    Select an action $a = \pi_\phi(s) + \psi$ where $\psi \sim \mathcal{N}(0, \sigma)$
    Execute a in the environment and observe the reward $r = \mathcal{R}(s, a)$, new state s' and done signal d
    Store the experience (s, a, r, s', d) to M
    $s \leftarrow s'$
  for $i=1,\ldots,K$ **do**
    Sample a mini-batch of N experiences $\{(s, a, r, s', d)\}$ from M
    $\tilde{a}' \leftarrow \pi_{\phi'}(s') + \psi$ where $\psi \sim \text{clip}(\mathcal{N}(0, \tilde{\sigma}), \alpha, -\alpha)$
    $y \leftarrow r + \gamma(1 - d)\min_{i\in\{1,2\}} Q_{\theta_i'}(s', \tilde{a}')$
    Update critics Q_{θ_i} to minimize $\frac{1}{N}\sum(y - Q_{\theta_i}(s,a))^2, \forall i \in \{1,2\}$
    if i mod *delay=0* **then**
      Update actor π_ϕ to maximize Eq. (2)
      $\theta_i' \leftarrow \tau\theta_i + (1 - \tau)\theta_i'$
      $\phi' \leftarrow \tau\phi + (1 - \tau)\phi'$

---

6 Simulator Design

To facilitate training and evaluating algorithms, we build an event-driven simulator to simulate order generation, battery changes and decision execution.

Data Description. The data provided by Cainiao include the orders in the Wuxi warehouse for 25 days. Each order contains the releasing time and the detailed job allocation for each working area. For each job, we can infer its duration via its start time and end time.

Timeline Design. We divide a day into 510 discrete time steps. During a time step, the following activities are executed sequentially.

- *Order generation.* Since orders in consecutive time steps are correlating in the real-world scenario, it is inappropriate to bootstrap from the real-world data for each time step individually. On the other hand, if we directly replay the real-world orders in a day, the number of different order pace patterns could be quite limited. As a result, the trained policy could overfit these patterns. Therefore, we propose to compromise by bootstrapping from long term patterns. Specifically, we divide a day into 24 slots (1 h per slot) and uniformly select an input for each slot from the real-world data in the same time interval.
- *Interacting with policy.* The policy computes an action according to the current state and submits it to the simulator. The simulator executes the numerical action by casting it to the set of AGVs to be scheduled and repositioning

these AGVs to their destinations (i.e., charging poles or working areas) with random delays.

- *Processing orders.* The simulator processes orders by a first-in first-out (FIFO) strategy. Specifically, it iterates over the order queue and attempts to allocate AGVs for the remaining jobs of each order. The allocation for an order terminates if (1) there is no available AGVs in the working areas or (2) the prepositional jobs haven't been fulfilled (e.g., the casting line job depends on the consolidation job). An order is considered as fulfilled and is removed from the order queue if its casting line job has been finished.
- *Updating AGVs.* For each AGV, the simulator updates its battery according to the duration of each status it has experienced in the time step. If its battery is lower than the dead threshold T_{dead}, then the AGV is marked as dead and cannot be scheduled to fulfill any job.

After finishing these procedures, the simulator returns a feature vector that represents the current state and a reward that is the number of orders fulfilled in current time step.

7   Empirical Evaluation

In this section, we conduct extensive empirical evaluation to demonstrate the effectiveness of our proposed method.

7.1   Experimental Configurations

We consider the problems with $|G| = 640$ AGVs, the charging capacity $C = 120$, the emergency charging threshold 30% and the dead threshold $T_{dead} = 15\%$. The size of the training instances ranges from 3,500 orders to 4,500 orders per day. For each episode, we randomly choose a training instance as order inputs. The test instances we consider are categorized into low loads (∼3,500 orders per day), median loads (∼3,900 orders per day) and high loads (∼4,300 orders per day). For each of configuration, we generate 50 instances and report the means as the results. The performance metrics we consider are listed as follows.

- *Fulfilled ratio.* The metric measures the proportion of the fulfilled orders at the end of an episode, which directly reflects the productivity of a policy.
- *Number of bottleneck time steps.* We consider a time step as a bottleneck time step when there is no available AGVs. Intuitively, if a policy fails to provide more available AGVs when the working areas are busy, then bottleneck time steps are more likely to happen, especially when considering the effect of backlogs.
- *Latency per order.* We consider the latency of an order as the duration required to fulfill the order. If a policy fails to provide enough AGVs, then the system has to wait for available AGVs and the latency of each order will increase.

– *Average number of idle AGVs.* The metric considers the number of idle AGVs in each time step. Intuitively, a poor policy would provide less available AGVs in each time step and thus the average number of idle AGVs is low.

The competitors we consider are three rule-based policies parameterized by $\langle T_c^t, T_w^t \rangle$ where T_c^t is the charging threshold for AGVs in working areas and T_w^t is the working threshold for charging AGVs in time step t. The fixed thresholds policy we consider has static thresholds $T_c^t = 40\%$ and $T_w^t = 80\%$, which is widely used in real-world scenarios. The dynamic charging threshold policy has a fixed working threshold $T_w^t = 80\%$ and a dynamic charging threshold T_c^t which is computed by an upper bound T_c^{UB}, a lower bound T_c^{LB} and the number of charging AGVs in current time step. Formally,

$$T_c^t = T_c^{UB} - (T_c^{UB} - T_c^{LB})\frac{|G_c^t|}{|G|}$$

In the experiments, we set $T_c^{UB} = 75\%$ and $T_c^{LB} = 35\%$. Finally, the dynamic working threshold policy has a fixed charging threshold $T_c^t = 40\%$ and a dynamic working threshold T_w^t which is computed by an upper bound T_w^{UB}, a lower bound T_w^{LB} and the number of working AGVs in current time step. Formally,

$$T_w^t = T_w^{UB} - (T_w^{UB} - T_w^{LB})\frac{|G_{working}^t|}{|G|}$$

where $G_{working}^t \subseteq G_w^t$ is the set of working AGVs. In our experiments, we set $T_w^{UB} = 80\%$ and $T_w^{LB} = 60\%$.

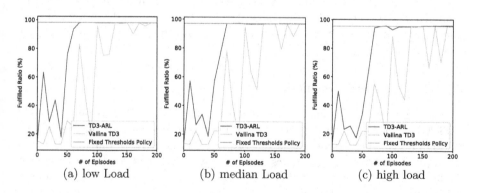

(a) low Load   (b) median Load   (c) high load

Fig. 3. Learning curves of TD3-ARL and Vanilla TD3 under different loads

For the implementation of the TD3 algorithm, both the actor and the critics are parameterized by a four layer fully-connection network where the hidden layers include a 300 neuron and a 400 neuron layers. Relu activation functions are applied to layers other than the last layer. For the output layer of the actor, a sigmoid activation function is applied to bound the action space. To represent

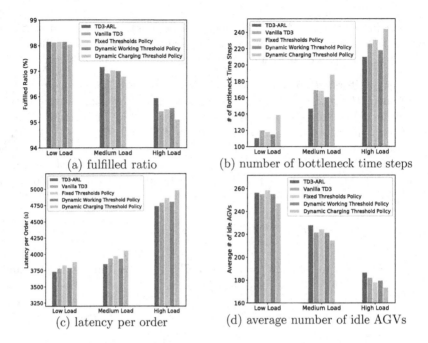

Fig. 4. Performance comparisons under different loads

the action component U_w efficiently, we squash the range $[|G| - C, |G|]$ to $[0,1]$ and the action can be directly recovered from the raw output of the actor via a linear transformation. Finally, we set the size of replay memory to 10^6, the standard deviation of exploration noise $\sigma = 0.1$, the standard deviation of target smooth noise $\tilde{\sigma} = 0.05$, the noise range $\alpha = 0.1$ and the policy update frequency to 2.

7.2   Experimental Results

Learning Curves. To evaluate the stabilities of vanilla TD3 and our proposed TD3-ARL, we test algorithms every 10 training episodes and present the fulfilled ratios in Fig. 3. It can be seen that our proposed TD3-ARL not only outperforms the vanilla TD3 but also improves much stably. In fact, TD3-ARL outperforms the fixed charging threshold policy before the 100-th episode and its performance still improves slowly afterward. On the other hand, the performance of vanilla TD3 oscillates in a broad range and it can only compete after 180 episodes. That is because ARL can effectively regulate the magnitude of an output action to guarantee the diversity of the experiences. In contrast, vanilla TD3 uses state independent noise to enforce exploration, which performs poorly due to the existence of equivalent actions and significantly slows down the learning process.

Performance Comparison. We benchmark the performance of different policies under different loads, and present results in Fig. 4. It can be seen from the

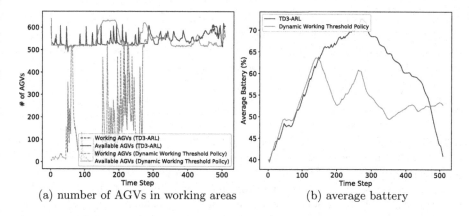

(a) number of AGVs in working areas   (b) average battery

Fig. 5. Behaviors of policies when solving the instance with high load

Fig. 4(a) that TD3-ARL can fulfill more orders than the rule-based policies, especially when solving high load instances. That is due to the fact that low load instances are relatively trivial that even a naive policy with emergency charging mechanism can easily fulfill most of orders. As the scale of instances grows, the rule-based policies no longer can compete as they cannot adaptively adjust their actions to proactively improve productivity. Figure 4(b) demonstrates the superiority of TD3-ARL over the rule-based policies in terms of the number of bottleneck time steps. Compared to the rule-based policies, our TD3-ARL can adjust the number of charging AGVs according to the current workload, and has less bottleneck time steps. Figure 4(c) shows the superiorities of TD3-ARL over the rule-based policies in terms of order latency. Although the dynamic working threshold policy considers the number of working AGVs and results in a lower latency than the ones of the other rule-based policies, it is still inferior to TD3-ARL, which highlights the effectiveness of proactive scheduling. Finally, Fig. 4(d) demonstrates our superiority in terms of idle AGVs. The results indicate that our TD3-ARL can provide more available AGVs than the rule-based policies, especially when solving the instances with high load.

Behavior Analysis. To look deeper into the decisions made by policies, we analyze the behaviors of TD3-ARL and dynamic working threshold policy on a high load instance and present results in Fig. 5. It can be seen that for the dynamic working threshold policy, the number of available AGVs in each time step is fairly stable. That is due to the fact that the policy is built upon energy conservation. As a result, it fails to take current workload into consideration and provide more available AGVs during the peak periods (i.e., after the 300-th time step). In fact, it schedules many AGVs to charge when the working areas are busy and can only provide about 520–540 available AGVs for working areas. In contrast, our TD3-ARL policy exhibits more deliberate behaviors, i.e., scheduling more AGVs to charge at the beginning of an episode (and the average battery increases) and scheduling more AGVs to work when the working areas are busy (and the average battery decreases). Besides, it is worth noting that in TD3-ARL the average battery at the end of the episode is close to the one at the beginning of the episode, which indicates that TD3-ARL is able to take the full advantage of the accumulated energy to improve productivity.

8 Conclusion

In this paper, we investigated battery management problem for large-scale automated warehouses which employ battery-powered AGVs to fulfill orders. To cope with its dynamic nature, we formulate the problem as an MDP with continuous state and action spaces. Since there are many equivalent actions in the MDP, traditional state independent exploration scheme performs poorly. Therefore, we propose a novel algorithm TD3-ARL which regulates the magnitude of the outputted action and enforces state dependent exploration via imposing a regulation loss term on the objective function of the actor. Extensive evaluations show the superiorities over the state-of-the-art, as well as the rule-based policies.

Acknowledgements. This work was supported by Alibaba Group through Alibaba Innovative Research (AIR) Program and Alibaba-NTU Joint Research Institute (JRI), Nanyang Technological University, Singapore.

References

1. Chou, P.W., Maturana, D., Scherer, S.: Improving stochastic policy gradients in continuous control with deep reinforcement learning using the beta distribution. In: ICML, pp. 834–843 (2017)
2. Coulom, R.: Efficient selectivity and backup operators in Monte-Carlo tree search. In: van den Herik, H.J., Ciancarini, P., Donkers, H.H.L.M.J. (eds.) CG 2006. LNCS, vol. 4630, pp. 72–83. Springer, Heidelberg (2007). https://doi.org/10.1007/978-3-540-75538-8_7
3. Ebben, M.: Logistic control in automated transportation networks. Ph.D. thesis, University of Twente (2001)
4. Enright, J.J., Wurman, P.R.: Optimization and coordinated autonomy in mobile fulfillment systems. In: Workshops at the Twenty-Fifth AAAI Conference on Artificial Intelligence (2011)
5. François-Lavet, V., Henderson, P., Islam, R., Bellemare, M.G., Pineau, J.: An introduction to deep reinforcement learning. Found. Trends Mach. Learn. 11(3–4), 219–354 (2018)
6. Fujimoto, S., Hoof, H., Meger, D.: Addressing function approximation error in actor-critic methods. In: ICML, pp. 1582–1591 (2018)
7. Haarnoja, T., Zhou, A., Abbeel, P., Levine, S.: Soft actor-critic: off-policy maximum entropy deep reinforcement learning with a stochastic actor. In: ICML, pp. 1856–1865 (2018)
8. Kocsis, L., Szepesvári, C.: Bandit based Monte-Carlo planning. In: Fürnkranz, J., Scheffer, T., Spiliopoulou, M. (eds.) ECML 2006. LNCS (LNAI), vol. 4212, pp. 282–293. Springer, Heidelberg (2006). https://doi.org/10.1007/11871842_29

9. Le-Anh, T., De Koster, M.: A review of design and control of automated guided vehicle systems. Eur. J. Oper. Res. **171**(1), 1–23 (2006)
10. Lillicrap, T.P., et al.: Continuous control with deep reinforcement learning. In: ICLR (2016)
11. McHANEY, R.: Modelling battery constraints in discrete event automated guided vehicle simulations. Int. J. Prod. Res. **33**(11), 3023–3040 (1995)
12. Mnih, V., et al.: Asynchronous methods for deep reinforcement learning. In: ICML, pp. 1928–1937 (2016)
13. Mnih, V., et al.: Human-level control through deep reinforcement learning. Nature **518**(7540), 529 (2015)
14. OpenAI: Openai five (2018). https://blog.openai.com/openai-five/
15. Rummery, G.A., Niranjan, M.: On-line Q-learning using connectionist systems. Tech. rep., Cambridge University (1994)
16. Silver, D., et al.: Mastering the game of go with deep neural networks and tree search. Nature **529**(7587), 484 (2016)
17. Silver, D., Lever, G., Heess, N., Degris, T., Wierstra, D., Riedmiller, M.A.: Deterministic policy gradient algorithms. In: ICML, pp. 387–395 (2014)
18. Silver, D., et al.: Mastering the game of go without human knowledge. Nature **550**(7676), 354 (2017)
19. Sutton, R.S.: Learning to predict by the methods of temporal differences. Mach. Learn. **3**, 9–44 (1988)
20. Watkins, C.J.C.H.: Learning from delayed rewards. Ph.D. thesis, King's College, Cambridge, UK (May 1989)
21. Zhao, M., Li, Z., An, B., Lu, H., Yang, Y., Chu, C.: Impression allocation for combating fraud in e-commerce via deep reinforcement learning with action norm penalty. In: IJCAI, pp. 3940–3946 (2018)
22. Zou, B., Xu, X., De Koster, R., et al.: Evaluating battery charging and swapping strategies in a robotic mobile fulfillment system. Eur. J. Oper. Res. **267**(2), 733–753 (2018)

Author Index

Printed in the United States
By Bookmasters